New Directions in Garden Tourism

New Directions in Garden Tourism

Richard W. Benfield

CABI is a trading name of CAB International

CABI	CABI
Nosworthy Way	WeWork
Wallingford	One Lincoln Street
Oxfordshire OX10 8DE	24th Floor
UK	Boston, MA 02111
	USA
Tel: +44 (0)1491 832111	
Fax: +44 (0)1491 833508	Tel: +1 (617)682-9015
E-mail: info@cabi.org	E-mail: cabi-nao@cabi.org
Website: www.cabi.org	

References to Internet websites (URLs) were accurate at the time of writing.

A catalogue record for this book is available from the British Library, London, UK.

Library of Congress Cataloging-in-Publication Data

Names: Benfield, Richard, author.
Title: New directions in garden tourism / by Richard Benfield.
Description: Wallingford, Oxfordshire, UK ; Boston, MA : CABI, [2021] | Includes bibliographical references and index. | Summary: "Following on from the success of Garden Tourism, this new book provides an update on the statistics and growth of the global phenomenon of garden visitation. It explores new themes and contemporary trends, from art and culture to psychographic profiling of visitors, and how social media and semiotics are used to enrich visitor experience"-- Provided by publisher.
Identifiers: LCCN 2020031738 (print) | LCCN 2020031739 (ebook) | ISBN 9781789241761 (hardback) | ISBN 9781789241754 (paperback) | ISBN 9781789241778 (ebook) | ISBN 9781789241785 (epub)
Subjects: LCSH: Tourism--Environmental aspects. | Garden tours.
Classification: LCC G156.5.E58 B455 2020 (print) | LCC G156.5.E58 (ebook) | DDC 338.4/791--dc23
LC record available at https://lccn.loc.gov/2020031738
LC ebook record available at https://lccn.loc.gov/2020031739

ISBN-13: 9781789241761 (Hardback)
       9781789241754 (Paperback)
       9781789241778 (ePDF)
       9781789241785 (ePub)

Commissioning Editor: Claire Parfitt
Editorial Assistant: Emma McCann
Production Editor: Tim Kapp

Typeset by SPi, Pondicherry, India
Printed and bound in the UK by Severn, Gloucester

Contents

Preface

_____

In 2013 I wrote the first book on garden tourism, entitled simply *Garden Tourism*. It was essentially a baseline summary of the nature, dimensions, current state, and prognosis for garden visiting as a form of tourism. It made the claim that garden visiting was the most popular form of contemporary outdoor leisure activity with a reach across all demographics, all continents, and all sectors of the tourism industry. My assertion has not changed in the 7 years since that was first written. Indeed, more and more evidence – statistical, anecdotal, and intuitive – has reinforced that belief. In 2018 the book publishers approached the author for an update. I indicated I was not interested in just updating the data or more dishonestly "slicing the ham a different way". It was not what they wanted. They pointed out that since 2013 gardens had grown in importance as tourism loci and that gardens had embraced many new technologies, marketing strategies, and physical improvements. There were also several new gardens and societal changes toward the plant world and food security and sources. They were correct. Moreover, the publisher pointed out that I taught a course entitled "New Directions in Tourism" at Central Connecticut State University (CCSU) and this would be a simple distillation of the material from that class. Not quite, but I was persuaded that indeed in the short 7 years since the previous edition, much had occurred that was worthy of separate treatment. This book describes in 60,000 words or so what has occurred.

To briefly illustrate, consider:

- More individuals visit gardens in the USA (120 million, more than twice the UK population) than go to Disneyland and Disney World combined; or that go to Las Vegas or Orlando, the USA's most popular tourism destinations.
- When asked what activity overseas visitors would like to pursue when they come to the UK, the third most popular activity, cited by 33%, is they would like to visit a garden, behind only "shopping" and "visiting a pub". In comparison with other activities undertaken by international tourists, visiting gardens was the most popular purely tourism activity, ahead of such activities as visiting museums, castles and religious buildings.
- In 2018 the National Trust welcomed 25 million visitors; that is one visitor for every household in the UK, and visitor growth to National Trust properties is currently over 5% per annum.
- Three gardens (Kew, Eden, and Chatsworth) rank in the top ten paid tourist attractions in the UK by number.
- An economic analysis in 2018 showed that in the UK, £2.2 billion could be attributable to overseas tourists visiting gardens and parks, or one-fifth of all tourism spending by overseas tourists.

- Gardens in the London metropolitan area, particularly Kew, Wisley, and even the once small Museum of Garden History across the Thames from the Houses of Parliament, were spending large sums of money – over £100 million – on visitor facilities and attractions.
- In May 2019 the UK Parliamentary Select Committee on Digital, Culture, Media and Sport convened a hearing on garden tourism because new research by VisitBritain had shown a dramatic rise in the volume, importance, and potential of garden tourism to stimulate visitation to the UK.

Clearly, since 2013, the numbers of garden tourists had grown along with their economic impact, new initiatives had been undertaken by many gardens, and garden tourism was one of the most dynamic and growing areas of tourism. New directions had certainly been charted by 2019 and thus a serious addendum to the material from 2013 was warranted. Here it is.

As is usual in a book of this nature, there are several disclaimers. The first is that the range and number of gardens referenced has diminished since the last edition. This is because the intent here is not to include every garden but rather to indicate generally what gardens are doing. Thus, not every garden is doing what I suggest is futuristic and moreover much of what they are doing is innovative and therefore often proprietary. It is appropriate that I acknowledge and thank those gardens who have contributed material and data from what is their own, often expensive, in-house or external research – many others have these types of data, but I was reluctant to ask them all to share. Therefore, I thank profusely those who did provide their data. As a result, as I say, I think the reader will find a limited number of gardens exemplified. Put more prosaically, if your favorite garden is not included it is *not* because I think them a poor example or that they were deliberately ignored or that they were uncooperative. The second disclaimer is that owing to the nature of new directions, many of the programs and initiatives of a garden might fit in many different chapters in this book. For example, the importance and ubiquity of festivals and events can be found in the festivals chapter, Chapter 7; as well as in the ethnic programs of urban gardens in Chapter 9 and new initiatives as part of the experience economy in Chapter 4. Third, I have deliberately tried to break the boundaries of simply discussing tourism to gardens. The tourism student would be better served by working beyond interdisciplinary boundaries and in the case of gardens moving into art history, historic structures, biology, urban geography, food security, and business product development and marketing; this book exposes students to some of these issues that are now part of 21st century garden tourism. Finally, this is the end; I am heading swiftly into retirement and it is time young researchers and academics expanded the field of garden tourism by means of quantification, observation, and challenging what I have now written (or researching areas I have not covered!). Once you have read the 60,000 or so words, I say, using the line from the poem *In Flanders Field* and repeated in the dressing room of the Montreal Canadiens: "To you from failing hands we throw the torch, be yours to hold it high."

In thanking those who have helped bring this book to fruition, first and foremost, I am grateful to all the garden tourism professionals who again made their knowledge and data freely available. Among the most valuable were Richard Deverell and Richard Barley at Kew; Mike Maunder, Becky Fisher, and Gordon Seagrave at Eden; and Huw Francis at the National Botanic Garden of Wales. Collectively these people are the best thing happening in garden tourism in the UK today. Sadly, Jo Connell at Exeter is now dedicating her remarkable talents to tourism events (of which garden tourism events are just one of many) and Dorothy Fox, the other seminal garden tourism author, has now retired. Yet again Andy Jasper, now at RHS Wisley, came through. Thanks to all the gardens of the American Public Gardens Association (APGA) and garden professionals who contributed to the book, principal among whom were Mary Pat Matheson and Sabina Carr at Atlanta Botanical Garden, Sheila Voss at the new Houston Botanic Garden, Keith Kaiser at Pittsburgh Botanic Garden, and Nick Leshi (again!) at New York Botanical Garden. Across the borough, Ronnit Bendavid-Val and Nina Browne of GreenBridge at Brooklyn Botanic Garden were most helpful. Adam Hill is doing the same sterling work as Nina in urban gardens in Philadelphia and was a wonderful resource. Ed Lyon at Reiman Gardens, Iowa told all on the astoundingly successful Lego Fest and the remarkable work being done in Miami, outlined in Chapter 9, emanates from the equally remarkable Amy Padolf at

Fairchild Tropical Botanic Garden. Thanks also go to Richard Mussler-Wright at Idaho Botanical Garden, William Tonks and John Graham at the State Botanical Garden of Georgia, Michelle Conklin at Tucson Botanical Gardens, Beth Monroe at Lewis Ginter Botanical Garden and finally Randy Fiveash, Connecticut State Tourism Director who is the nation's branding expert and whose input in Chapter 3 is obvious. Of all gardens, perhaps the most support and encouragement came from Peter Wyse Jackson and Liz Fathman at the oldest public botanic garden in the USA, Missouri Botanical Garden in St. Louis, Missouri ... with maybe the exception of old friend Susan Lacerte and her staff in Queens Botanical Garden in ... well, Queens, New York. Casey Sclar and his staff at the APGA were, as usual, supportive and helpful.

In Canada, Michel Gauthier is a driving force behind garden tourism around the world and he always tried to advance not only garden tourism but also my own garden tourism agenda. I think *he* is Mr. Garden Tourism. Lee Foote in Edmonton gave all the information on his new Islamic garden. In Victoria, Dale Ryan and Dave Cowen at The Butchart Gardens were, as usual, very helpful; much of the insight on social media comes from their astute use and analysis of social media. Jonah Holland at Lewis Ginter Botanical Garden and Johanna Dominguez of Put a Plant on It in Buffalo also brought an aging author into the new world of social media.

In the event field, my dear friend and fellow author Jim Charlier[1] and his partner in garden tourism crime in Buffalo, Elizabeth Licata, were great supporters. Also in Buffalo, New York, Gordon Ballard and Brian Olinski serve excellent beer alongside their spectacular Hemerocallis, even when not on Garden Walk Buffalo. Grace Elton, Rob Burgess, and the rest of the staff at Tower Hill Botanic Garden in Worcester, Massachusetts, were always helpful. Tower Hill is my closest botanic garden and hence received the brunt of my requests.

In the chapter on gardens and historic properties, the wonderful Kathrine Malone-France at the National Register of Historic Places not only told me about new directions in historic places but set the stage for wonderful visits and conversations with Howard Zar at Lyndhurst Mansion and Greg Sages and Cole Akers at the Glass House in New Canaan. It stimulated further work, results, and good times with Sherry Hack and Robert Lyon at Connecticut Antiquities. Not insignificant for any book, the beautiful cover is at the instigation, brow beating, and finally creation of the talented Kara Newport at Filoli Historic House & Garden, a national historic place.

In doing a book of this nature, one often comes across people in other areas or disciplines that spark admiration (and not a little envy) because of their talents. In the writing of this book, Alan McKeon at Alexander Babbage/TruTrade was not only helpful but taught me tons about 21st century technology and marketing and all it cost me was a few beers. Julia Blau at CCSU taught me all about Levy walk analysis.

Many thanks are due to Ginger McCurdy and her two siblings who loaned me their house and gave me access to their fridge full of Hawaiian beer, on Oahu, to finish the final edit which also allowed me to meet Joslyn Sand at the Hawaii botanic gardens authority and Jon Letman at the National Tropical Botanical Garden on Kauai.

My colleagues at CCSU helped immensely with the book's production. The unassuming Phil Meng created all the thematic maps of garden visitor zip codes while Brian Sommers and Tim Garceau tactfully and accurately edited the chapter on urban garden tourism. Diane Cannata supplied coffee, got our student workers to do my bidding, and provided deflection of incoming students and trouble (often one and the same person) with unfailing good humor.

Thanks go to Claire Parfitt, Ali Thompson and Tim Kapp at CABI who displayed amazing patience and tolerance toward a lazy and tardy writer, and Gill Watling who did an amazing job of copy editing.

Finally, thanks to my neighbors, all of whom think I have the best garden in the world, and Katherine Baldwin who travelled with me to many gardens and who knew I do not even have the best garden on the block.

---

[1] A shameless plug for the excellent book, *Buffalo-Style Gardens*.

Gardens by the Bay, Singapore. Opened in 2012, the 54-hectare garden now attracts 50 million visitors yearly and is a key part of making Singapore the "City in a Garden". Photo courtesy of Pascal Garbe and used with permission.

1

Introduction: Philosophy of *New Directions in Garden Tourism*

© Richard W. Benfield 2021. *New Directions in Garden Tourism* (R. Benfield)
DOI: 10.1079/9781789241761.0001

In 1954, the marketing guru Peter Drucker said:

> If we want to know what a business is we have to start with its purpose ... There is only one valid definition of business purpose: to create a customer. It is the customer who determines what a business is. For it is the customer, and he (*sic*) alone, who through being willing to pay for a good or service, convert's economic resources into wealth, things into goods. What a business thinks it produces is not of the first importance – especially not to the future of the business and to its success. What the customer thinks he or she is buying, what he or she considers 'values', is decisive. Because it is its purpose to create a customer, any business enterprise has two – and only these two – basic functions: marketing and innovation.
>
> (Drucker, 1954)

That paragraph summarizes, in a nutshell, what this book is all about. In 2013, the first edition of the book *Garden Tourism* was published. It essentially put between two covers what the history, significance, nature, and dimensions of garden tourism were. It was the first book to deal entirely with garden tourism, but it encapsulated why garden tourism was such an important part of both the tourism industry and the botanical world of gardens and garden visitation. It has now been 7 years since that first edition. In those 7 years garden tourism has grown remarkably, some would say exploded, and the innovations and marketing so fundamental to Peter Drucker, 60 years ago, have been realized, such that today garden tourism may be the most important and largest sector of contemporary outdoor leisure.[1] For this second edition, it would have been easy and desirable to just update progress in the field of garden tourism since 2013 but that would have left unaddressed the vast strides that have occurred in garden tourism in the past 7 years. It is the aim of this book to go beyond the baseline description of garden tourism developed in the first edition and examine the innovations and marketing in garden tourism that have made it what it is today.

The Current State of Research in Garden Tourism

While evidence exists that garden visiting is now one of the most important sectors of the leisure industry, it is lamentable that the academic research community has not embraced garden tourism. Indeed, in a review of the first edition, the noted tourism researcher Kevin Markwell (2014) commented "that there has been surprisingly little scholarly research conducted on garden tourism". In May 2017, as part of a Kew Gardens' conference on garden tourism, one of the few researchers over the years to recognize and address garden tourism undertook a content analysis of all English-language, peer-reviewed journals and articles with the words *garden* plus *visitor*, *tourist*, or *tourism* in the abstract (Fox, 2017a). She found 32 articles in 23 different journals (Fig. 1.1). The majority of articles were published in the USA (8), the UK (5), and China (4). Eleven other countries were represented in the literature (Fig. 1.2). Most (14) discussed botanic gardens, followed by urban parks and gardens (8) and some historic gardens (4) (Fig. 1.3). Most (66%) were quantitative in methodology and 22% were qualitative in nature (Fig. 1.4). The most topical areas were attracting visitors (Lee *et al.*, 2010; Byun and Jang, 2015), management of the garden (Banks, 2015), and new technologies (Pérez-Sanagustín *et al.*, 2016). In short, for the most popular contemporary leisure activity in the world, the attention of researchers and practitioners has been lamentable. In the first edition, it was suggested that garden tourism presents itself as a fruitful (and enjoyable) area of research. Sadly, in the years since, that challenge has not been taken up. As a result, readers will find, as material in this book, much that is a narrative of what has occurred in the years since 2013 rather than new academic insight and conclusions. Much of the narrative is based on individual gardens' own research or perspective; little is external. This is an area that must change if a more complete story of the tourism industry is to be told. To that end other areas neglected in the first edition have been addressed.

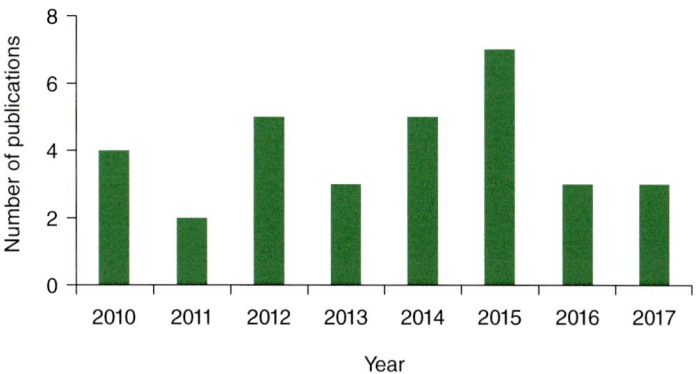

Fig. 1.1. Number of published articles on garden tourism, 2010–2017. Adapted from Fox (2017a).

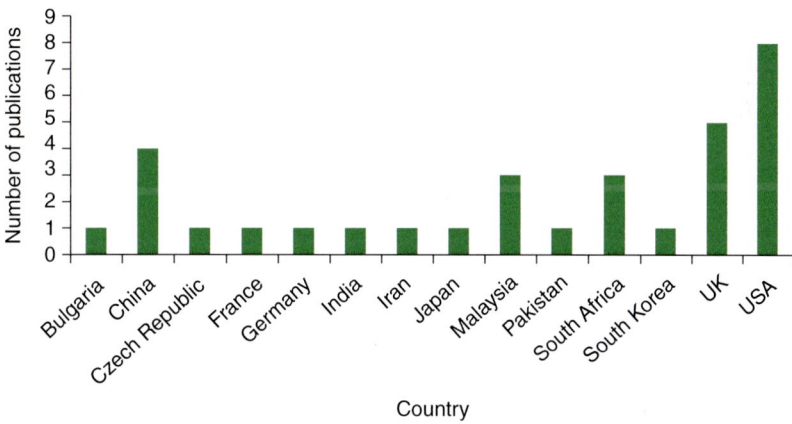

Fig. 1.2. Source of published articles on garden tourism, 2010–2017. Adapted from Fox (2017a).

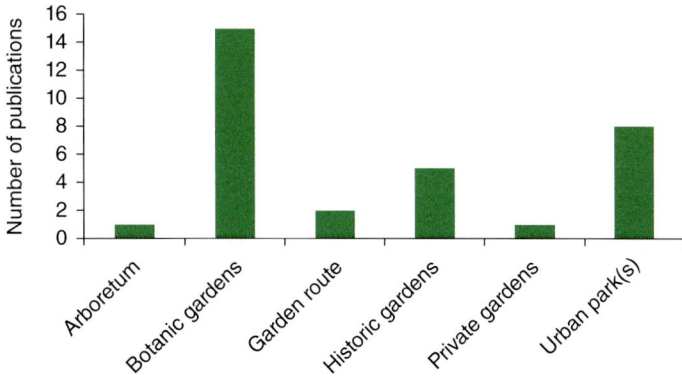

Fig. 1.3. Topic of published articles on garden tourism, 2010–2017. Adapted from Fox (2017a).

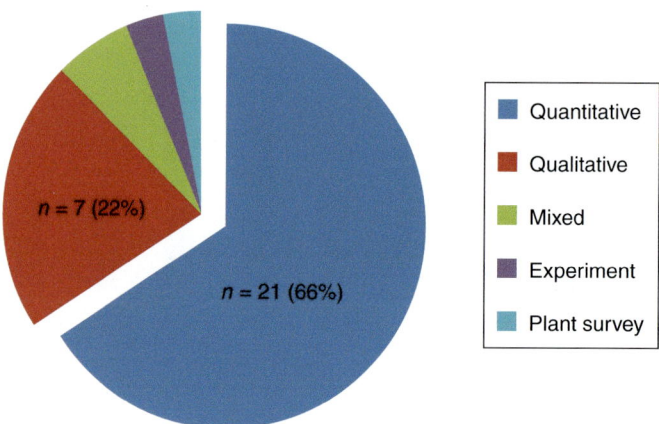

Fig. 1.4. Type of research conducted in published articles on garden tourism, 2010–2017. Adapted from Fox (2017a).

Markwell also pointed out in his perceptive review that:

> Readers will not find engagement with debates concerning authenticity, commodification, embodiment, emotions or the experience economy, nor does the book intersect with the literature on 'natures', constructed or otherwise. Yet gardens are surely one of the best examples of nature–culture hybrids. In this regard, Benfield misses an opportunity to examine how garden tourism provides opportunities for our understandings of the relationships between nature and culture to be deepened and, perhaps, challenged. Certainly, he does examine relationships between garden tourism and cultural tourism (29) and art (54) but he does not extend this examination into an engagement with deeper philosophical or theoretical concerns. I was also surprised at the lack of an extended discussion of gardens as significant elements of destination identity and branding. Again, I think there were opportunities to explore, in much more depth, the role of large public or commercially operated gardens in the creation of a destination brand or image for certain destinations.
>
> (Markwell, 2014)

These were valid observations and this new edition of the book addresses them in part; only in part because, as was noted earlier, there has been so little published in the garden tourism literature in areas such as authenticity, commodification, and branding. In summary, this book will update the reader on progress in garden tourism by introducing new gardens, new audiences, new strategies, new initiatives, and new programs that in total over the past 7 years are making garden tourism the dominant mode of contemporary outdoor tourism.

Innovation and Marketing; The Structure of the Book

In the following chapter, a selected appraisal of garden tourism around the world in terms of garden numbers, number of visitors, and growth is given. It updates numbers and development of garden tourism and contains a brief appraisal of how garden tourism has grown, or not grown, in the prior 5 years.

Chapter 3 examines new directions in garden tourism (innovation) by selecting seven major product development and marketing innovations that have characterized gardens in the preceding 7 years. They are:

- gardens and wildlife;
- art and gardens;
- gardens and music;
- Levy walk analysis and gardens;
- plant societies and gardens;
- sensory experiences at gardens; and
- garden branding.

Chapter 4 examines new audiences and particularly the greater sophistication in market segmentation since 2013. The demographer

David Foot claims that demographics explains "two thirds of everything"; hence changing demographics, 7 years on, is examined and particularly the growing numbers of, and interest in, the so-called "Millennials". If demography explains two-thirds of everything then perhaps the field of psychographics covers all or part of the remaining third. Increasingly gardens are using psychographic studies to segment their audiences and examples will be drawn from Kew and the National Trust in the UK, and Newfields in Indianapolis, USA. This chapter will also introduce two other market research developments in gardens that may be used for segmentation bases. These are Claritas and geofencing, both of which are inherently geographical in their base of analysis.

Chapter 5 examines the new media landscape of communities (blogging and podcasting), platforms, and social media, a phenomenon that has exploded in the past 7 years such that Al Reis[2] says "in marketing today there is social media and everything else".

Chapter 6 explores the deep psychological draw and attraction of flowers, gardens, and gardening, first examined under the study of semiotics by John Urry in the late 20th century and which is still little examined as a cause or motivation for garden visiting.

Chapter 7 is dedicated to the study of garden-related events. Getz and Page (2016a,b) chart the phenomenal rise of event tourism and the concurrent rise in academic research. More specifically, Connell *et al.* (2015) highlight the use of events to fill gaps left in off-peak times and seasons and it is in this area that gardens have developed significant programs and events to make them more financially stable and educationally viable.

Chapter 8 goes into more depth than the previous edition on the economics of garden tourism because significantly more gardens are examining the economic benefits of their gardens. As important, the chapter also highlights the environmental, health, and social benefits of gardens in an era of environmental sustainability, and social justice.

Chapter 9 focuses on the urban environment in which garden tourism takes place. Many gardens are now characterized by meaningful, visual, and impactful outreach into both the surrounding communities and the city in which they are located. In some cases, the crossover into streetscaping and events have become significant generators of urban garden tourism. As a result, one of the examples this book uses – the Chelsea Fringe Festival in London and elsewhere – was featured in Chapter 7.

Chapter 10 isolates historic garden tourism and the current management and development of historic gardens as opposed to the purely historic nature of gardens developed in the previous edition. Here the focus is on landscape, which is a more holistic approach to historic sites, and marks innovation in historic garden tourism.

In Chapter 11, the future of garden tourism is examined. Moskwa and Crilley (2012) indicate that botanic gardens have multiple roles but principally education, environmental, and recreation, and it is under these three roles that garden tourism's future will be evaluated.

The final chapter is essentially a postscript written as the effects of Covid-19 changed, so dramatically, the nature of garden operation and visitation. Much of the impact is drawn from the American Public Gardens Association (APGA) surveys of its member gardens in March (the immediate effects) and April 2020 (the results of the closure on revenue, staffing, and programs). In the final paragraphs the measures being taken to permit a (partial) reopening and the effects of the partial opening are examined.

New Gardens

In 2004 Connell charted the rise of gardens in the last hundred or so years, suggesting that the development cycle has been strong and certainly not at a stage of decline. Hulme (2015) charts the rise of botanic gardens over time for the six continents and shows the particular rise of garden establishment since 1950.[3] In fact, the growth of gardens and especially gardens open to the public has been nothing short of spectacular in the last 20 years. Woods (2018) has produced a listing of the most spectacular gardens over the last 20 years in his book *Gardenlust*. It highlights, in pictures and in a horticultural and design narrative, over 100 gardens built worldwide only in the last 20 years.

In the field of product development, the attraction of building new gardens for tourist and botanical purposes has not yet reached the end. The APGA receives and admits applications for at least ten gardens annually. As this is being written, four gardens in the USA are close to opening for the first time: Pittsburgh Botanic Garden (the city's second after the iconic Phipps Conservatory), Pennsylvania; Gardens on Spring Creek, Fort Collins, Colorado; Delaware Botanic Gardens at Pepper Creek, Dagsboro; and Santa Fe Botanical Garden, New Mexico. At least three others are in the planning stage including Houston, Texas,[4] which does not have a botanic garden.

In Asia it is estimated that five new botanic gardens are opened each year in China, while in the Middle East, the progress of opening gardens so evident 5 years earlier has abated. In Australia, Botanic Gardens Australia and New Zealand (BGANZ) admits approximately one new garden annually to join the 75 already in existence.

New Audiences

Concurrent with new gardens, new audiences have been attracted to gardens. Much has been made of the rise of the Millennials, those people born between 1981 and 1996 and now reaching family and household formation stages. Much of the discussion over the preceding 5 years has been consumed with the desire to attract this generation, in part because a decline in Baby Boomer numbers and interest in gardens has been forecast. Chapter 4 will suggest Baby Boomer interest in gardens and garden visiting has not waned – some data suggest it might have increased – while Millennials have been attracted to gardens and garden visiting in increasing numbers to the delight of garden managers the world over. David Foot, the world renowned demographer, suggests that Millennials were never a threat to abandon gardens and the like; rather they are doing, in household formation, exactly what the Baby Boomers did some 20 years earlier – they are buying (suburban) houses, making or inheriting gardens, and undertaking activities just as their parents did a generation earlier.[5] In total it can be seen from the data that since 2013, which is when the effects of the recession of 2008/09 had effectively ended, garden visitation exhibited one of the highest and most significant growth rates in the tourism industry (growing at 7% per annum), ranking in the same league as cruising (growing at 4%), visiting amusement parks (growing at 4%), and general sightseeing (percentage unknown).

Existing Gardens; New Initiatives, New Uses

In the previous edition, reference was made to new gardens in which the new garden was failing or not enjoying the growth initially anticipated, desired, or expected from a new product. Two significant examples were highlighted, the National Botanic Garden of Wales in Carmarthenshire, Wales and the Eden Project in Cornwall, England. In the 7 years since the first edition both gardens have proceeded to gain significant attendance numbers such that Eden now has over 1.2 million visitors, up from 859,000, and the National Botanic Garden of Wales has become one of the finest gardens in the UK and especially in its contribution to Welsh plant and animal conservation.

Alnwick Castle – garden revival

Alnwick Castle is one of the iconic castles in Northumberland, UK, that have guarded the North Sea for over 700 years. The garden, which has existed since 1759 and was an original Capability Brown-designed landscape garden, was in a state of complete disrepair by the turn of the century. The Duchess of Northumberland wanted to not only revive the garden but also add contemporary features to it to make it a modern tourist attraction. Redevelopment commenced in 2001 and in 2004 it opened to the public. The garden is described as "a huge public garden of classic symmetry and astonishing beauty..." that is not only an inspiring landscape garden but also contains water features for guest interactions, learning, and play.[6] Like many gardens, seasonal attractions are used to bring year-round visitation. Thus spring is the time of bulbs, in summer the ornamental garden is in full bloom and the rose garden contains over 3000 David Austin roses, autumn is renowned for the only pleached crab apple trees in the UK, and the year concludes with frozen cascade waterfalls, a light show, and lantern parades. Alnwick Castle received a significant marketing boost in 2000 when the first in the series of *Harry Potter* films was shot in and around the castle. Today, the garden boasts over 350,000 visitors per year.

Alnwick Castle – the Poison Garden

Billed as "The Most Dangerous Garden in the World", in 2005 Alnwick Castle and gardens installed a Poison Garden. With restricted entry and a separate entry fee, it is possibly the only garden in the world where smelling, tasting, or even touching is strictly forbidden. The Poison Garden is visited by 90,000 of the 350,000 annual visitors. The garden contains over 100 plants both deadly and intoxicating and while there is a decided thrill about entering the garden, the plants are also used as a part of a drug education program and as a public service for plants that are dangerous.

National Botanic Garden of Wales

The National Botanic Garden of Wales is now a leader in research investigating which plants are the most important to pollinators and disseminating these findings to farmers and the gardening public (see Case Study 11.1).

RHS Wisley

The UK has 27 million gardeners out of a total population of 64 million. More precisely, over 85% of UK households have a garden and the popularity of gardening is expected to increase well into the 2020s, notwithstanding an increasing number of houses in the UK without a garden.[7] Much of the expertise and contact with the gardening industry in the UK has been through the Royal Horticultural Society (RHS). Established in 1804 as the Horticultural Society of London, it acquired a site at Wisley in 1903 and from that time the RHS has been offering gardening advice to 60 million enquiries per year and running five garden sites as examples of gardens and plants dedicated to the furtherance of all things garden.[8]

The five RHS gardens have over 2 million visitors per year with the garden at Wisley having 1,071,000 visitors in 2018, placing it in the top 20 of all paid attractions in the UK. The RHS is a charity or not-for-profit, and its work is aimed at the 490,000 members and 30,000 schools in the country. As a major future direction, it dedicated £160 million in 2015 to further "enrich everyone's life through plants and make the UK a greener and more beautiful place".

The investment projects are:

1. At RHS Wisley, build a new horticultural science and learning center, build a new welcome center, and restore Wisley village.
2. Build a fifth display and educational garden at Bridgewater, Salford, Manchester.
3. At the RHS Garden Harlow Carr, restore the historic Harrogate Arms and bath house as well as the landscape.
4. At the RHS Garden Hyde Hall, build a new visitor center as well as a new learning center, activities, and food-service building.
5. At the RHS Lindley Library, modernize the access to the collection.
6. Develop new programs for the RHS to provide community outreach and establish urban gardens, thus bringing gardens to the cities.
7. Raise the profile and opportunities for careers in horticulture.

The majority of the investment monies was dedicated to the flagship garden at Wisley. The welcome center provides a much-needed upgrade to the existing entry, it will include a new restaurant and shop, and bring the plant center into the complex. The welcome center complex has been designed for open space meetings and in conjunction with the display gardens to provide a spectacular entrance to the garden.

The second major fixed-roof structure will be a new national center for horticultural science and learning. The science will be centered on horticultural science, taxonomy, plant health, and plant diseases. All these activities will be observable by the visitor and integrated into the educational portion of the building. Here the facilities include meeting rooms/convention facilities, an advisory center for the UK's gardeners, and three outside educational gardens related to daily living. As Wisley describes it, the center's facilities will deliver and engage gardeners with four key themes:

1. Global knowledge bank on gardening and garden plants.
2. Plant health in gardens.
3. Gardens in a changing world, particularly one undergoing climate change.
4. Plant science for all people, plants, and the planet.

It is anticipated that the new facilities and orientation will increase visitation to over 1.5 million annually and if this were realized, Wisley would rise into the top ten paid tourist attractions in the nation.

The Eden Project

In March 2001, the Eden Project in Bodelva, Cornwall opened and by June 2001, over 1 million people had visited. Numbers in the first year were over 1.5 million visitors. In that opening year Eden was in the top ten tourist attractions in the UK and was a major prospect for tourism growth in the coming years. Visitation declined dramatically in the subsequent years, such that by 2013 numbers had dropped to 859,000 – a drop of 45% – and it was no longer in the top 30 tourist attractions in the nation. Something clearly had to be done to arrest this decline. In 2014 a fivefold strategy was developed to address this decline. These elements were:

1. To better understand the visitor and why they come to Eden. At the most basic, it was delineated that there was a market potential of 1.4 million visitors to Cornwall but of that group only 162,000 came who were dedicated visitors with high interest in horticulture. Clearly there was potential to increase numbers from this group.[9] Research showed that of that group 86% were holidaymakers, 49% had never been to Eden before, and their propensity to visit was shown by the fact that 30% of visitors visited more than six other gardens per year. Clearly, this was a lucrative segment to reach and, for this group, a dedicated campaign called the "Great Outdoors" was initiated. To attract this group changes were made in the Eden product at the request of gardeners. Specifically, 80%

requested increased labelling of plants, 55% requested a dedicated garden guide, and 50% indicated a preference for an introductory talk and tour.
2. Market segments. What this told Eden was that garden visitation was essentially split into two groups:
    a. Locals who constituted 80% of the visitors in winter.
    b. Out-of-town visitors who constituted 80% of visitors in summer (20% were local).
Marketing strategies were aligned to reflect this division; thus locals could be persuaded to undertake repeat visits (probably in the range of 20,000 or more local visitors) and memberships to the garden at £20 could be increased to this market segment from the current 75,000.
3. Marketing changes. It was perceived that the current marketing was scattered and thus too diluted. As a result, marketing in national publications was discontinued and in particular, marketing in the London Metropolitan area ceased while more emphasis was placed on the local market, recognized in point 2 above. As one Eden marketing executive indicated, "we began to fish where the fish were".
4. Digital marketing. Eden, much like other gardens, saw the importance and effect of digital marketing and thus embarked upon a strong presence in social media, particularly on Facebook (see Chapter 5).
5. Finally, and probably most importantly, Eden embarked on a series of Live Programming events to give visitors a reason to come. The number of events went from 12 in 2014 to 25 in 2018.
    a. Starting in 2016 programs were built around space and space exploration and ran for two years. In 2019, this changed to a focus on biodiversity and extinctions.
    b. In 2016, a display of dinosaurs but with an emphasis on education and science was installed not only for the UK school holidays but also extending into the summer months.
    c. A PRIDE festival was held in 2016, and an October marathon and half marathon (2010) were greatly expanded in the years since 2016.
    d. The Eden Classic bike ride was started in 2014 and consists not only of a 100-mile circuit of Cornwall, but also features foods and culinary displays. A BMX display featuring

Matt Hemmings[10] was part of the event focus and he gave a cycling display and exhibition aimed specifically at a youth audience.

e. Other events included a Cornish pasty baking contest and art displays.

f. Finally the largest and most ambitious event, Christmas in Cornwall, ran throughout November and December, and with lights, choristers, Father Christmas, and a dedicated skating rink, Eden was positioned to be the destination for Christmas holidays, so much so that three national newspapers placed it in "the UK's best places for Christmas this winter" in 2018.

The results of these new directions have been remarkable. By 2017 visitor numbers had again surpassed 1 million and the Eden Project was the 12th most popular paid tourist attraction in the UK. In 2018 Eden received 1,006,928 visitors, putting it into the top ten of the UK's paid visitor attractions.

Garden hotels

Particularly in the UK, many castles, palaces, monasteries, mansions, and country seats have been notable for their gardens, both in size and historical importance. But after 1945, many properties became financially unsustainable when death duties and inheritance taxes forced the conversion of many houses to other uses and were converted to trusts, or leased properties, or were sold off or donated to the National Trust. Many of these properties have evolved into historic properties and their importance for garden tourism is covered in Chapter 10. However, a large number were converted to hotels and thus the marriage of upscale tourism and hospitality properties (particularly hotels but also conference centers and restaurants) with gardens has become an important new direction. In Germany, the most famous garden hotel is the Anholt Wasserburg in Anholt, Rhine Westphalia; in India, the iconic Udaipur Lake Palace in Udaipur is a particularly famous hotel with somewhat smaller and restricted gardens; while many villas in Italy (Villa Cipressi, Villa D'Este on Lake Como, and Villa Serbelloni) are both hotels and destination gardens. Garden hotels also exist in Spain, Iran, Portugal, and even Guatemala. In the UK, Cliveden House in Buckinghamshire with its famous formal gardens is a hotel, conference center, and wedding venue, most recently the location for Meghan Markle's stay on the night before her marriage to Prince Harry. More restrained is Gravetye Manor, the former home of the originator of the English country garden, William Robinson. Today it has 17 rooms, accommodating 35,000 guests per annum either for the accommodation and/or restaurant. The property is open only to hotel guests but can be visited by special arrangement, much like Rosemary Verey's garden, Barnsley House in Gloucestershire, England.

Gardens and literature

The link between gardens and literature is a strong one, and far too voluminous to detail here. The two literary gardens mentioned most often[11] are the Garden of Eden in Milton's *Paradise Lost*, Book IV and the most famous children's book, Burnett's *The Secret Garden*. However earlier medieval literature is replete with gardens, including but not limited to de Meun's *Romance of the Rose*, Marie de France's *Lais*, Machaut's *The Judgment of the King of Bohemia*, plus Chaucer's *Book of the Duchess*, and *The Knight's Tale* and *The Merchant's Tale* in *The Canterbury Tales*. In Spenser's *The Faerie Queene*, the Garden of Adonis is found in Book 3, canto 6, and especially the "Bower of Bliss" at the end of Book 2.

Of course, Shakespeare features the garden and many plant species – perhaps most uniquely in Richard II where the gardeners compare tending a garden to running a kingdom! In 18th century literature, Rochester has a rose garden in *Jane Eyre*, and Andrew Marvell's "The Garden" and especially "The Mower against Gardens" are seminal poems of this century. Examples in 19th century literature are the Pyncheon Garden in *The House of the Seven Gables*; Hawthorne's garden in "The Old Manse" given to him and Sophia by Thoreau as a wedding present; and Poe's "The Domain of Arnheim" is essentially a treatise on esthetics that uses landscape gardening for its premise. All the March girls have their own little gardens in *Little Women*. The garden in Hawthorne's "Rappaccini's Daughter" is famous now having been replicated as the Poison Garden at Alnwick Castle. Gardens are quite dynamic in Lewis Carroll's *Alice's Adventures in Wonderland* and *Through the Looking Glass and What Alice Found There*. Gardens figure heavily, too, in several of Oscar Wilde's fairy tales in the collection *The Happy Prince and Other Tales* and also in "The Nightingale and the Rose", "The Selfish Giant", and "The Devoted Friend".

Many of the gardens of English literature still exist for the tourist to visit. Most notable is Shakespeare's garden at Stratford-upon-Avon. The walled garden of Great Maytham Hall, inspiration for *The Secret Garden*, still exists and is open as part of the UK National Garden Scheme. The Shakespeare Garden at the Brooklyn Botanic Garden showcases the plants noted in the collected works of Shakespeare and for poets, Wordsworth's house, Dove Cottage and garden at Grasmere in the Lake District, close to Beatrix Potter's – or, more accurately, Mr MacGregor's – garden at Hill Top, is open for tourist visits.[12]

In the realm of poetry, just about ANY poem written in the 17–18th century in England features a garden! John Denham's "Cooper's Hill", Yeats's "Lake Isle of Innisfree", Coleridge's "This Lime Tree Bower, My Prison", Keats's "Ode To a Nightingale", and the Song of Solomon 4.12–16.

One of the more interesting gardens with direct tourism links to poetry are the 37 botanical gardens throughout Japan dedicated to exhibiting the plants referenced in the 8th century *Man'yoshu* poetry anthology. This anthology contains approximately 4500 poems, of which 1600 refer to one or more of about 160 different species of plants, making botanical references and imagery one of the most significant features of the work. Despite the chronological distance separating the *Man'yoshu* from the present day, there are 37 dedicated Man'yo botanical gardens in Japan that vary widely in type: some are part of larger public parks, some are attached to shrines or temples, some are attached to museums of various types, and some are independent, but all are clearly intended to act as facilities to attract visitors to localities or institutions. They are, therefore, stimulators of literary contents garden tourism. By being dedicated to the objects referenced in poetry, the gardens form an unusual type of contents tourism facility, which is focused on the resources of, and stimulants for, literature, rather than the literary work itself (see McAuley, 2016).

Case Study 1.1: Missouri Botanical Garden – How One Garden is Combining New Initiatives

Events

Realizing that many potential visitors work during the day when the garden is open, one of the trends in event planning is for more evening events. At Missouri Botanical Garden in St. Louis, this ranges from months-long events like the summer of 2019 Garden Party Lights and followed in the autumn/winter with the popular winter holiday show, Garden Glow, to one-night pop-up evening hours that allow visitors to enjoy the garden after work and see some of the flowering plants at peak bloom time (cherry trees and azaleas were two that were successfully tried in 2019). After-hours drinking events are popular, and Missouri Botanical Garden hosts an annual wine tasting (Grapes in the Garden), beer tasting (Fest-of-Ale), and local spirits tasting (Spirits in the Garden – held near Halloween). In addition to the popular Whitaker Music Festival, the garden added more evening music events at satellite gardens in other regional locations. Finally, the garden offers several nighttime hikes at the nature reserve. The hope is to continue to offer pop-up evening hours and other evening events to take advantage of the light quality at dusk, when the garden is at its most beautiful.

In addition to the challenge of the garden's hours competing with the workday, like all gardens Missouri Botanical Garden has to contend with inclement weather. The garden is, by nature (pun intended), an outdoor venue, and people have become less inclined to venture out when the weather is even the slightest bit bad. On one occasion the garden decided to "lean in" to the weather and on a day where the temperature forecast for the St. Louis area was in the vicinity of zero degrees all day, the management opened the garden for free all day and invited people to come visit the tropical Climatron to escape the frigid temperatures. In this spirit the garden continues to look for other opportunities to "lean in" to bad weather (e.g. espousing that kids + rain = fun!) rather than apologize for it.

Services at the garden

The revamped William T. Kemper Center for Home Gardening has a new plant doctor and information desk providing one-stop help for home gardeners. Missouri Botanical Garden believes it is the only one of its kind in North America and possibly the world. The garden is also partnering with a local farmers' market to have a regular booth there to answer gardening questions, especially during planting and harvesting seasons. Finally, the garden offers a sustainability hotline where people can get advice about green and sustainable living and business practices.

Social media

As of June 28, 2019, Missouri Botanical Garden had 171,000 fans on Facebook, 49,000 followers on Twitter, and 64,000 followers on Instagram. The latter is its fastest-growing social media platform and is well suited to hosting stunning pictures of the garden's plants and events. Like many gardens, it encourages engagement through use of event-specific hashtags and occasional photo contests with prizes (usually an annual membership). Instagram is also the source of many of the garden's influencers, with whom the garden engages to help spread the word about the garden. In 2017 the garden started a blog called Discover + Share.[13] The year ended with 27,500 views, and 2018 ended with 55,000. At the time of writing, 2019 will end with over 60,000 views. For this initiative, the most popular posts have been about trees, garden history, and gardening help.

Community outreach

This is considered by the garden to be a significant growth area although not a new direction. Missouri Botanical Garden believes it has a great message but for a long time kept the message hidden under a basket, so to speak. The garden was a founding member of a regional initiative called BiodiverseCity St. Louis,[14] a network of organizations and individuals throughout the greater St. Louis region who share a stake in improving quality of life for all through actions that welcome nature into their urban, suburban, and rural communities. In addition, the garden is involved in a host of youth programs, community partnerships, and regional initiatives where it can offer expertise in biodiversity conservation and sustainability. The garden has also greatly expanded its therapeutic horticulture program both in the garden and out in the community, partnering with local hospitals, shelters, a crisis nursery, and eldercare facilities to bring the healing power of plants to patients and practitioners alike.

Finally, in late 2019 Missouri Botanical Garden engaged the services of a branding agency (see Chapter 5 on branding) to help hone its message and tighten the grasp on its brand.

Notes

[1] Statistical Abstracts of the USA (2012) suggest gardening is the second most popular leisure activity after "walking".

[2] Chairman, Reis and Reis. Credited with coining the term "positioning" in the field of marketing.

[3] Hulme was examining the status of botanic gardens only, not display or other types of garden, and was concerned only with the role they play in conservation and/or pathways for plant invasions.

[4] Planned opening Fall 2020.

[5] D. Foot, Toronto, 2019, personal communication.

[6] In this regard Alnwick Castle harkens back to the pleasure water gardens of the 17th century (see Heilbronn Garden, Salzburg, Austria).

[7] Over 2 million residences in the UK do not have gardens and this number is increasing.

[8] The RHS also runs many garden shows, the most famous being the Chelsea Flower Show, and it also coordinates Britain in Bloom.

[9] Research also showed that those with a high interest in horticulture spent more time at Eden, spent more money during their visit, were more likely to recommend Eden to others and also become repeat visitors, and they gave higher satisfaction and "value for money" scores than other groups.

[10] In the USA this would be the equivalent of Tony Hawk riding and performing.

[11] As gardens as a focus or stage for literature are far too voluminous, I asked the excellent faculty of Central Connecticut State University's English department, all of whom possess a singular and significant area of expertise, to name their top literature examples featuring gardens. This is that list and I am grateful to each of them for their suggestions and contributions.

[12] With a dedicated Peter Rabbit garden.

[13] Available at: https://discoverandshare.org (accessed July 24, 2020).

[14] Available at: https://www.missouribotanicalgarden.org/sustainability/sustainability/biodiversecity-st.-louis.aspx (accessed July 24, 2020).

Chihuly Garden and art under the Space Needle, Seattle. Reproduced by kind permission of the author.

David Austin Rose Garden, Cheshire, UK. Reproduced by kind permission of David Austin Rose Garden.

Kew Broad Walk, 2015. Author's own photo.

Kew Broad Walk, 2016, after new design by Richard Wilford. Author's own photo.

Prior to entry into the Poison Garden at Alnwick, Northumberland, one is required to register by QR code for possible Covid-19 tracing. Dean at the entrance will also monitor social distancing as well as giving excellent tour information. Author's own photo.

Changi Airport, Singapore. Garden between terminals. Photo courtesy of Singapore Airports Authority.

Jardin Botánico Canario Viera y Clavijo, 7 km from Las Palmas de Gran Canaria, is home to over 500 plant species endemic to the Canary Islands. Author's own photo.

2

Gardens Around the World, 2013–2019

———————

© Richard W. Benfield 2021. *New Directions in Garden Tourism* (R. Benfield)
DOI: 10.1079/9781789241761.0002

While it is not the intent to just update the data from 2013, the fact that garden visitation around the world is growing at such an astounding rate is worthy of regional analysis and explanation. Remarkably, in the UK, garden visitation has been estimated as growing at approximately 7.8% per annum and in the USA, growth is probably in the 5% range. This growth seems present both in domestic tourism and international arrivals, pointing to a demand for gardens as a tourist destination by all nationalities and age groups.

The USA

With over 570 gardens as members[1] of the APGA, the USA along with the UK represents the highest number of gardens in the world. Few states have no public garden and many, like the Philadelphia region, boasts "30 gardens within [a radius of] 30 [miles]". Furthermore, gardens are being added to the inventory at a rate of five per year and the programs and events at gardens in the USA suggest they are some of the most innovative in showing new directions in garden tourism. Furthermore, with the undertaking of a national benchmarking study in 2014/15 (for the year 2015) that is now ongoing and delivering data on a garden's size, location, employees, volunteers, and visitation, US gardens are becoming some of the most strategic at defining management and operations in the context of tourist visitation and, of course, conservation.

The data compiled by the APGA are impressive. Over 120 million people are estimated to have visited a public garden in 2018,[2] while over 5 million K–12 students (2.3 million from Grades 1–5) participated in an education program. Median volunteer hours reported was 1850, while 4 million visitors participated in conservation programming. As businesses, public gardens in the USA are major enterprises with over 70 having an annual operating budget above US$3 million and over 20 having annual operating budgets above US$10 million. With the figure of 120 million visitors, gardens rank among the most popular leisure activities, surpassing visits to Las Vegas, Orlando,[3] and Disneyland and Disney World combined, and far ahead of the world's cruise passengers.[4] It may not be an overstatement to say that in the USA garden visiting is the most popular contemporary outdoor tourist activity!

As examples of innovation and tourism focus, some of the most successful have been installations at gardens to attract both seasonal and new audiences. At Longwood Gardens, Pennsylvania, the USA's pre-eminent display garden, a Bruce Munroe series of installations in 2012 was the forerunner to a number of art installations that in many ways acted as substitute for the fountains and lights show that was closed between 2014 and 2017 for a US$90 million renovation. Today the garden attracts over 1.5 million visitors with a wide array of events, special performances, and installations that supplement some annual attractions like A Longwood Christmas, the chrysanthemum show following Thanksgiving, and summer performances. While Longwood may be the most spectacular and noteworthy of America's garden destinations, smaller gardens have developed interesting and successful additions to the garden that have increased visitation. The Berner Botanic Garden in Des Moines, Iowa was the first garden to feature LEGO© that has subsequently been exhibited in five other botanic gardens in the US Midwest.[5]

Canada

Canada, with only 48 or so public gardens, and less than 15% of North American tourism arrivals, has made significant attempts to develop a garden tourism industry. Butchart Gardens in Victoria, British Columbia is considered one of the most iconic gardens in the world with almost 1 million visitors annually but, as important, it has aligned itself with a number of other gardens on Vancouver Island to promote a garden route for garden tourists, in the belief that garden tourists will participate in a number of garden visits as part of a tourism journey. Canada also boasts a Garden Tourism Coalition dedicated to advancing garden tourism in the country. The coalition has, as its members, display gardens like Butchart, indigenous people's gardens, historic gardens, and classical botanic gardens like the Royal Botanic Garden in Hamilton, Ontario, the Toronto Botanic Garden, and Montreal Botanic Garden. Perhaps the most striking new garden development in the whole of North America was the opening in 2018 of the Islamic Garden as part of the Devonian Botanic Garden (now the University of Alberta Botanic Garden) in Edmonton, Alberta. Built with US$25 million in funds given as a gift by the Aga Khan, the garden immediately went from a historic average of 70,000 visitors per year to over 130,000 in 2018 (Fig. 2.1).

Mexico

There are 21 public botanic gardens in Mexico. The largest is the Garden of the Botanical Institute at the Autonomous University of Mexico, Mexico City, attracting 222,000 visitors per annum. The second and third largest are in

Fig. 2.1. University of Alberta Botanic Garden. Photo courtesy of Paul Swanson.

Sinaloa, the most westerly state of Mexico and containing the city of Mazatlán. For tourism purposes, the gardens in the Yucatán Peninsula (Merida and Chetumal) and Mazatlán have become very important as possible locations for significant international tourist visitation. Those gardens in the Yucatán have been slow to recover from recent hurricanes but the garden in Vallarta, which was established in 2004 and by 2019 was attracting 40,000 visitors annually, is not only growing rapidly in visitor numbers but is also acquiring a reputation as one of the best gardens in the world.

South America

In contrast to much of the world, garden and garden tourism development in South America is moribund. Brazil with 43 gardens has the most of any country in South America, with the iconic Botanical Garden in Rio de Janeiro the most famous and most visited, but after Rio, Brazilian gardens are generally poorly funded and thus facing an uncertain future. Peru and Argentina have some noteworthy gardens but many, such as in Lima and Buenos Aires, are de facto city parks in the center of the urban area. Colombia has 20 gardens listed as botanic gardens, with the garden in the capital, Bogotá, hosting over 300,000 visitors per year. As a new direction in tourism it was the first location in Latin America to host, in January 2013, SonEra Solar – the first solar-powered music festival in Latin America, a group effort between the garden and the US Agency for International Development, which enabled the garden to have the first festival of its kind where the instruments and equipment were powered completely by solar power. Venezuelan botanic gardens are currently in a poor state owing to the turmoil racking the country as this is being written.

The possible exception to this sad state in South America is the Quito Botanical Garden, Ecuador. Only open since 2005, the garden is in the center of Quito and attracts over 80,000 visitors per year. In an attempt to grow and become relevant, the garden management has embarked on an ambitious and impactful program to make the garden sustainable. Ecuador also contains, in the municipality of Tulkan, a cemetery, which as a cemetery may be unremarkable but contains over 300 topiary structures of various animals and geometric shapes.

The UK – England, Wales and Northern Ireland

A recent (2018) report by Oxford Economics described the UK as "the gardening capital of the world". By extension it may be that the UK is also the garden tourism capital of the world. With 33% of all UK international visitors (11.9 million) expressing a desire to visit gardens when they come to the UK, garden visiting is the most popular outdoor activity overseas tourists participate in during their visit to the UK. If one adds to that the over 57 million domestic or day visitors to gardens and the 10 million overnight domestic garden visits, one gets a total of over 70 million visitors to gardens in the UK, the highest per capita garden visitors in the world. This astounding number owes its origin to a long and diverse garden history but more recently the garden community in the UK has embarked on a number of noteworthy and significant initiatives that have propelled garden tourism into the most important attraction for tourists. Given the large number of gardens and garden attractions in the UK it would be impossible to cover or categorize these initiatives, but some significant examples stand out as follows.

Royal Botanic Gardens, Kew, Richmond, Surrey

Following the recession of 2009/10, Kew Gardens experienced a dramatic decrease in tourist visitation from historic highs of around 1.8 million visitors per year to 1,023,000 visitors in 2012, a decrease of over a third. Coupled with declining or a threatened decline in government support, Kew embarked upon a dedicated policy to reach out to new audiences and increase the number of events for visitors to attend, rather than just passive viewing of the gardens. Both of the aforementioned are described elsewhere but three of the most important new products were the opening of the Children's Garden in 2019, the refurbishment of the Temperate

House, completed in 2018, and the installation of a long walk in 2016, an avenue of annual and perennial plants leading from the main gate to the glasshouses. Of note is that these garden beds represent the first exterior garden installation since possibly the 1950s or even earlier! The result of these and the other attractions targeted to new audiences has been an increase in visitation to over 1.8 million visitors in 2017 and 2018, despite inclement weather in those years. Preliminary figures from 2019, and projections into future years based on anticipated long-term success of the new temperate house and the opening of a children's garden, could possibly create visitor numbers over 2 million, making it the fourth-largest paid attraction in the UK.

Chester Zoo and botanic garden

In 2016, Chester Zoo attracted almost 1.9 million visitors making it more popular than any other garden in the UK. Certainly, the fact that it contains animal species as well as plant species would seem to inflate the numbers, but other zoos with similar contents do not approach these numbers and it seems there are remarkable management strategies at work to stimulate visitation that should be explored.

National Trust

While there are very few dedicated botanic gardens in the UK, the National Trust, the registered charity, has 215 gardens listed in its 2017 Handbook. It is the major destination, along with English Heritage properties,[6] for UK garden visitors. Visitor numbers to National Trust properties have grown from just 270,000 in the 1970s to an historic high of 26.6 million in 2018, and the number of members joining the Trust has rocketed by over 1 million in just 5 years to hit a record 5.2 million, according to the charity's 2018 annual report. Of concern to the Trust is the fact that the overall proportion of visitors that rated their visit as "excellent" fell from 67% in 2016/17 to 61% in 2017/18. Similarly, those visitors who rated their experience as "very enjoyable" fell from 66% to 56% between 2013/14 and 2016/17.

Much of this decline has been attributable to food-service provision, parking, and toilet facilities, and thus over the next 5 years the Trust is expending considerable funds to upgrade all three areas.

National Garden Scheme

In 2019 the National Garden Scheme hosted 631,000 visitors in 3552 gardens. However, the National Garden Scheme not only provides garden visits and promotes the scheme, it has also expanded its impact on the role of gardens and gardening in society by commissioning, in 2016, a 65-page report on the effects of gardens and gardening. While it does not specifically address the impact of garden visiting on health and well-being, the report does recognize the importance of gardens as a leisure activity away from the house, stating "We also visit gardens as part of visiting historic sites (40.2 per cent) or taking 'days out' (68.7 per cent)", and indicates the link between gardening, domestic travel, and in particular historic sites (Fig. 2.2).

Scotland

Perhaps owing to the importance and the image of Scotland as a location for historic castles and houses, gardens have taken a subsidiary role in the product offerings in Scotland. The most visited garden in Scotland is the Royal Botanic Garden Edinburgh (RBGE) with 929,140 visitors in 2018, up 2.5% from the previous year, and it is certainly the most progressive garden in Scotland for new directions and programs. The Insight Department of VisitScotland, Scotland's national tourist organization, has researched the overseas market to gauge the propensity of overseas holiday visitors to visit gardens.[7] Among all overseas visitors to Scotland, visiting gardens is third in popularity with 56% of visitors, after "castles" and "famous monuments" (e.g. Edinburgh Castle).

Much like RBGE, 44% of all Scottish garden visitors are locals, 26% other Scottish, 25% other UK (mostly England), and 6% overseas. For UK visitors in particular over 40% have never been to Scotland and get their information on Scotland from online travel sites (53%), guidebooks and brochures (43%), the national tourism website (42%), and other travel websites and blogs (34%). Thus, it appears that the English tourist is heavily influenced by computer-based information. This is borne out by data that find 43% of English visitors spend over 20 hours per week on the Internet and that 69% are active on Facebook, 29% on Twitter, and 25% on Instagram. At least 70% go on Facebook daily and 50% on Twitter. Finally, of the English garden visitors, they are predominantly male (59%!), usually without children (66%), and highly educated. Perhaps most surprisingly,

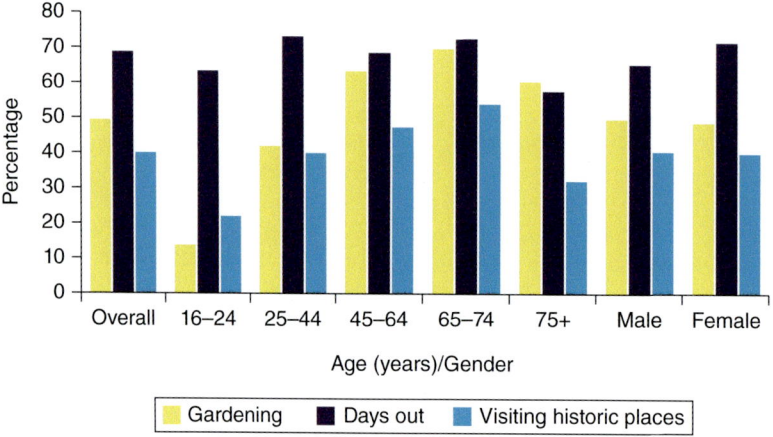

Fig. 2.2. Percentage of adults reporting gardening, visiting historic places and days out as free-time activities in 2013/14. Adapted from Buck (2016).

of those English visitors who say garden visiting is the main purpose of this trip, of those aged 25–34 years, 37% say the garden visit is the main purpose of their visit (as opposed to 17% of all tourists), suggesting that garden visitation is NOT just for elderly females. Similarly, for those aged 55 years or older, 22% say visiting a garden is the main purpose of their trip, again higher than the percentage for total British tourism.

Unlike in England, Scottish National Trust properties are not major locations for gardens, rather they are usually historic homes and castles. The most popular of the National Trust for Scotland's 38 gardens are Inverewe Garden (191,951 visitors) and Pitmedden Garden with 43,045 visitors in 2017. In total, in 2017, Scottish National Trust gardens recorded over 1 million visits. More important than the Scottish National Trust gardens are the private gardens open for Scotland's Garden Scheme. In 2019 there were over 500 gardens open, of which 52 were new to the scheme. It is thought that more than 50,000 visitors went to gardens open under this scheme in 2019. Many gardens in Scotland are unique in the gardening world and rank among some of the most important gardens globally for design, thus attracting significant numbers of visitors. The Garden of Cosmic Speculation in Dumfries is only open for one day each year but will attract over 5000 visitors on a fine day.

Ireland

Ireland may be one of the most progressive garden tourism destinations in the world today in its embrace of gardens not only as a generator of tourism revenue, but also its use of gardens in branding its regional destinations. The tourism challenge for Ireland was that the main attraction (and gateway) of Dublin drew in the largest number of tourists, with the west coast of the country, for its scenic and spectacular attractions, being the main attraction once visitors left Dublin. Thus in 2017 Fáilte Ireland focused on the gardens and historic houses of Ireland's Ancient East to lengthen the tourism season, add new experiences, increase spending, and disperse visitors throughout the island. The marketing plan was named the "Great Houses and Gardens Plan" and consists of product development, events, and new experiences based on the garden product. The potential for this program to succeed certainly exists, with three of the top ten paid attractions in Ireland being houses and gardens with over half a million visitors each.

Europe

There are probably around 3000 gardens in Europe, making it the region with the highest concentration of botanic gardens in the world.[8] The 27 members of the EU plus Switzerland, Slovenia, and Iceland contain 527 purely botanic gardens which are closely affiliated as part of the European Botanic Gardens Consortium. Fig. 2.3 and Table 2.1 show the distribution by country of those gardens in the Consortium, suggesting an asymmetry in garden distribution but more seriously an asymmetry in gardens located in areas of endemic or threatened plant species. Northern Europe contains 288 botanic gardens or approximately 17% of the world's gardens but the Mediterranean region, one of the world's biological hotspots containing most of Europe's endemic or threatened plants, has far fewer botanic gardens.

Fox and Edwards (2008) indicated that France was as important a garden tourist destination as the UK and the 2018 Avion Train Route Survey confirmed this fact with 38% of all visitors declaring they had visited a park, garden, natural site, or forest, which translates to about 19,434,000 visitors.

Official data from the France Tourist Development Agency in 2018 confirm this number, indicating that 25.2% of French domestic tourists (7,026,000) undertook a visit to a garden park or natural area while 53.9% (11,952,000) of international tourists visited a garden. Table 2.2 indicates the dimension of visitation to a sample of gardens in France.

While the countries and gardens of Eastern Europe have been integrated into the European network of gardens since 1991, they are significantly different from Western European gardens. Most of the botanic gardens in Central or Eastern Europe are legacies of the socialist era and are thus located within university grounds and administrative structures. Hungary, for example, has 22 botanic gardens, all located in, or associated with, universities and other institutes of higher education. In contrast, gardens in Italy, Switzerland, and

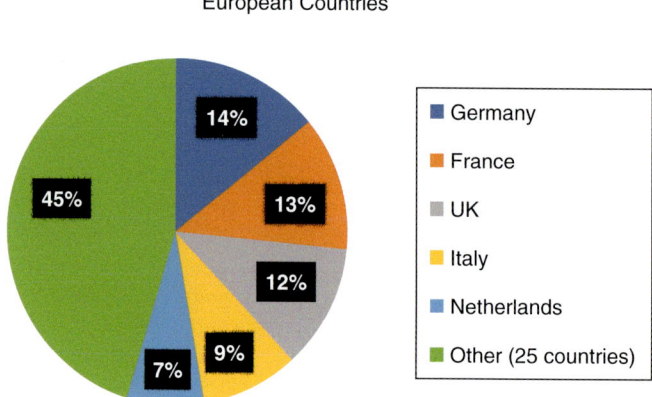

Fig. 2.3. Distribution of botanic gardens in 30 European countries, 2015.

Table 2.1. Number of botanic gardens in selected European countries, 2015.

Country	No. of botanic gardens	Remarks
Germany	74	The German Association of Botanic Gardens has 97 member institutions
France	66	Only 30 gardens (2017) were members of the national network Jardins botaniques de France et des pays francophones
UK	61	There is no botanic garden network in the UK, but gardens do cooperate in plant exchanges
Italy	48	See below and Italian Botanical Heritage
Netherlands	39	
Greece	4	

Austria represent some of the most iconic gardens in terms of scenic setting, visitor numbers, and garden history. The Italian Botanical Heritage guide to Italian gardens lists 2800 parks, display gardens, historic gardens, and other specialized gardens,[9] while Italian alpine gardens are some of the most famous in the world, and yet in this region gardens are still being built as tourist attractions. The Gardens of Trauttmansdorff Castle in Merano, Italy are less than 20 years old but have won travel awards and competitions, hosting almost half a million visitors per year.

Cameroon, Gabon, and Togo still suffer from inadequate funding and a generally depressed tourism performance in world tourism patterns. Much of the work in African gardens is focused on plant conservation, food security, and the medicinal use of plants. One notable development has been the establishment (opened in 2012) of a new garden in Addis Ababa, Ethiopia and some significant botanical successes in Tanzania and Reunion. Most typical and probably the most visited as African gardens are the botanic gardens in Entebbe, Uganda[10] and Nairobi, Kenya.

Sub-Saharan Africa

Little has occurred in new directions in garden tourism for African gardens, except for South Africa. The West African gardens of Ghana,

South Africa

While South Africa's share of tourism arrivals to Africa has decreased somewhat in recent years, at 16% of arrivals and 23% of tourism receipts,

Table 2.2. Attendance at selected French gardens, 2018.

Garden	No. of visitors[a]	Increase over previous year (%)	Remarks/location
Jardin des Plantes (Ménagerie)	507,000	2.0	The menagerie of animals is attached to the free garden. Paris
Parc zoologique de Paris	537,000	18.5	
Parcs zoologiques Lumigny	350,000	11.6	
Zoo du bois d'Attily	85,000	63.5	
Parc zoologique de Thiory	450,000		
Ecole nationale supérieure de paysage – Portager du Roi (Versialles)	20,000	−1.8	
Arboretum de Versailles–Chevreloup	24,000	75.6	
Domaine de Chamaronde	120,000		
Parc de la Propriété de Caillebot	124,000		
Domain départemental de Montauger	20,000		
Parc et Château Domaine de Courances	24,000	11.5	
Domaine départemental de la Vallée-aux-Loups – Arboretum	221,000	12.8	
Domaine départemental de la Vallée-aux-Loups – Jardin de l'Ile verte	33,000		
Roseraie du Val-de-Marne	20,000	−16.8	
Giverny	696,556		Monet's garden. Location 40 miles from Paris
Château de Villandry	360,000		Possibly the most famous French chateau garden

[a]As most gardens are free, accurate numbers are difficult to obtain.

South Africa is a major force in tourism to African gardens; in 2018, there were 10,472,000 visitor arrivals to South Africa. South Africa's ten national botanic gardens registered 1,327,189 visitors in the 2018 calendar year (domestic and international) and in the 20 years between the end of apartheid in 1994 and 2014, South Africa's national botanical gardens hosted 22 million visitors – a rise of 155% during the 20-year period. This is due, in large part, not only to some iconic gardens (Kirstenbosch, Cape Town, which hosted over 13 million of these visitors), but also a national network of gardens all under the management of the South African National Biodiversity Institute (SANBI), which has been proactive in building new gardens[11] as well as developing tourism strategies to justify and support their existence and funding. The future of South African gardens has been summed up in South Africa's National Gardens Expansion Strategy: 2019–2030 (Willis, 2019), which, while addressing the biological need for South Africa to conserve the remarkable and unique diversity and importance of its nine biomes, suggests that its gardens can be expanded through both the establishment of new gardens and the expansion of existing gardens. At the time of writing, new gardens have been proposed for the Limpopo Province (Thohoyandou) and processes are underway to find a suitable site for a new national botanical garden in the North West Province, thus ultimately providing a national botanical garden in all of South Africa's nine provinces. This book would be remiss if it did not mention the contribution and example set by Kirstenbosch National Botanical Garden when, in April 2014, it opened the "boomslang" (literally meaning "tree snake"), an aerial walkway running 12 m above the ground and 130 m in length set entirely within and above the tree canopy. As Figs 2.4 and 2.5 indicate, visitation rose dramatically.

Australia

Australian gardens are represented by the 110 member gardens of BGANZ of which seven

Visitors (including free entry, excluding concerts)

	Apr	May	Jun	Jul	Aug	Sep	Oct	Nov	Dec	Jan	Feb	Mar	Financial year to date
Year 09/10	48,581	33,533	23,717	35,802	39,447	40,479	64,786	56,479	73,759	57,972	55,582	64,053	594,190
Year 10/11	56,955	36,379	25,469	33,340	38,950	53,253	60,547	58,076	75,754	63,501	56,452	58,353	617,029
Year 11/12	56,404	35,663	26,301	43,174	40,176	44,469	63,136	62,194	77,832	68,667	61,081	62,488	641,585
Year 12/13	48,164	36,883	26,407	31,236	34,569	51,944	63,772	70,745	81,771	74,579	63,516	71,033	654,619
Year 13/14	55,905	44,885	33,960	35,599	42,780	51,270	63,980	70,077	89,196	74,286	67,079	67,432	696,449
Year 14/15	68,770	47,884	58,551	63,139	75,297	77,390	96,270	86,117					573,418

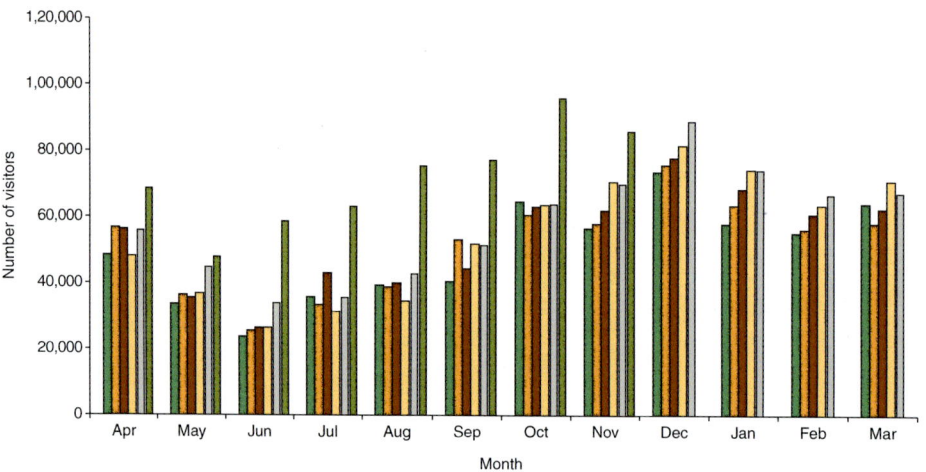

Fig. 2.4. Visitors to Kirstenbosch National Botanical Garden, Cape Town, South Africa, 2009–2015. Reproduced courtesy of SANBI.

Gate income (excluding concerts)

	Apr	May	Jun	Jul	Aug	Sep	Oct	Nov	Dec	Jan	Feb	Mar	Financial year to date
Year 09/10	898,326	564,844	349,446	575,130	742,794	813,551	1,424,415	1,308,028	1,598,730	1,438,681	1,468,740	1,479,988	12,662,673
Year 10/11	1,065,064	775,223	443,346	541,574	743,243	1,060,056	1,466,513	1,464,597	1,755,382	1,618,758	1,447,916	1,409,316	13,790,988
Year 11/12	1,224,228	716,829	515,800	716,315	772,113	990,874	1,545,587	1,675,719	1,978,340	1,967,560	1,585,494	1,532,895	15,221,754
Year 12/13	1,149,256	729,052	568,216	622,218	753,737	1,211,424	1,821,028	1,750,795	2,071,151	1,818,425	1,615,264	1,736,974	15,847,540
Year 13/14	1,199,196	941511	594,847	680,492	970,578	1,262,402	1,657,365	1,970,380	2,431,164	2,026,703	1,897,647	1,877,104	17,509,389
Year 14/15	1,791,657	1,161,350	1,423,944	1,272,528	2,060,876	2,249,980	2,898,720	2,722,770					17,581,825

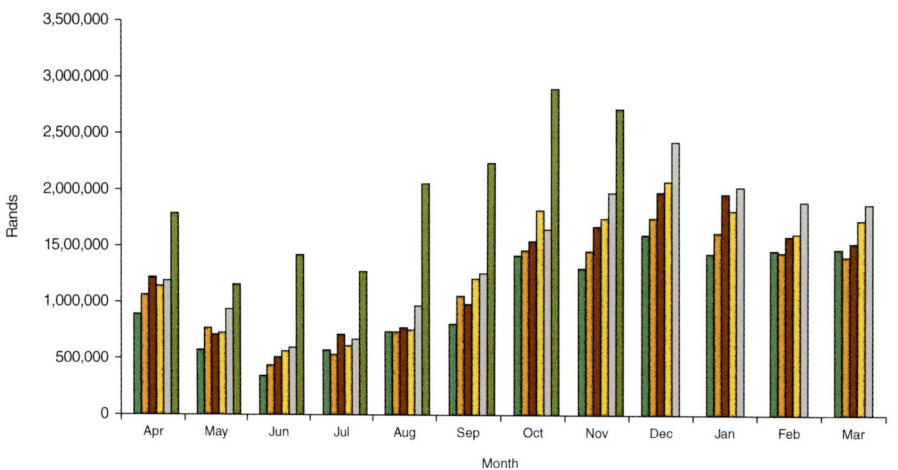

Fig. 2.5. Gate income, Kirstenbosch National Botanical Garden, Cape Town, South Africa, 2009–2015. Reproduced courtesy of SANBI.

members are in New Zealand and the rest in Australia. Approximately two new member gardens join each year but there are also a number of small member gardens that participate in BGANZ activities. In 2019 one of the new gardens to join BGANZ was Inala Jurassic Garden in Tasmania, a dedicated educational garden focusing on plants and connections with the ancient continent of Gondwanaland and establishing the links of plants before and after continental drift. With a population of over 28 million, Australia is well represented in terms of garden product, with every major city in the country having what may be considered an iconic garden and all of the smaller mid-sized cities having a botanic garden that is generally well supported and provides a major leisure resource for Australians. Visitation numbers on the gardens have not been updated since 2013 but data from some individual gardens would suggest visitation is growing at about 5% per year.

New Zealand

Garden tourism in New Zealand is primarily run through the New Zealand Garden Trust (NZGT),[12] a consortium of 120 gardens throughout the two islands. Most major cities have a botanic garden that is free and attracts significant numbers of visitors. However, the most remarkable is Hamilton Gardens, some 2 hours south of Auckland, that claims to be the most visited tourist attraction in New Zealand and in fact has won international garden tourism awards. The 54-hectare site contains 21 "themed" gardens with a strategic development plan to build representative gardens of the major historic garden periods. Currently Hamilton Gardens plans to complete four "new" gardens in the next four years. Thus, the most recent gardens were the Mansfield Garden, the Concept Garden, and the Chinoiserie Garden. There will be the fourth garden in the series after completion of the other three.

For New Zealand as a whole in 2013, "visiting botanic gardens" was the seventh most popular activity of international visitors which represented 243,000 visitors or 18% of all visits. The majority of garden visitors come from

Australia (19% of Australians visited a garden), but proportionately the British (21%), German (20%), and Chinese visitors (19%) had a greater propensity to visit gardens than Australians. Interestingly, garden visitors stayed longer in New Zealand and spent more per visitor compared with other tourist activities. As is the case in many, if not most, national tourism promotion agencies, notwithstanding the importance, value, and numbers of garden visitors, gardens as a separate category in tourism product offerings is subsumed under tourist activities that are certainly as ubiquitous as in other countries (golf, shopping) and certainly attract fewer tourists than gardens.

East Asia, including China

China is one of the most biologically rich countries in the world for plant diversity and has a high level of endemism. However, population growth, industrialization, and urbanization threaten much of this diversity. Heywood (2017) estimates 20% of China's higher plants are threatened with extinction. The response by the Chinese Communist Party has been dramatic. From just 34 in 1960, China now has over 180 botanic gardens but is adding as many as five gardens per year in very different biomes.[13] The gardens are primarily research institutions mostly operated by the Chinese Academy of Sciences and dedicated to plant conservation. Thus tourism is not a major focus of the gardens but, even so, some of the gardens post impressive numbers. Xiamen Botanical Garden has over 1 million visitors per year, and Kunming Botanical Garden and Xishuangbanna Tropical Botanical Garden[14] in the remote west and south respectively host 400,000 and 600,000 visitors annually. Shanghai Chenshan Botanical Garden that opened in in 2011, 50 miles from Shanghai, now accommodates over 1.1 million visitors with a wide range of unique gardens and educational facilities and Ningbo Botanical Garden that opened in 2016 posted 700,000 visitors in its first year. More telling perhaps for their suitability for tourism is the estimate by He and Chen (2012) that only 20 of the country's botanic gardens have visitor education centers.

West Asia and the Gulf States

It is generally agreed that the Middle East is lacking in the requisite number of gardens to represent the large number of plant species in the arid region. In 2013 there was hope that this would be rectified with plans for botanic gardens throughout the region in such places as Sharjah, Al Ain, and Dubai in the UAE, and Qatar. The plans were never realized and the reason for the lack of follow-through was the Arab Spring of 2010–2012 that generated retrenchment by the ruling families in the Gulf States particularly.[15] At the time of writing, the Oman Botanic Garden is still in the process of being completed, with the highlight being a large glass structure capable of exhibiting all of the biomes in the country of Oman. Completion date is unknown. The Jordanian Royal Botanic Garden founded in 2005 has yet to open to visitors, but planting, facility development, and a research agenda are being actively pursued. The singular exception to this pattern of very slow development realization is the Dubai Miracle Garden. Started in 2013 and entering its seventh season, the garden attracts over 1.5 million visitors in the November–May period during which it is open. The garden is purely a flower garden,[16] being redesigned every year with large floral displays and exhibits in various types of floral arrangements. In doing so the garden claims many a "first". In 2018 it received the award for the largest topiary and in 2016 was recognized by *The Guinness Book of World Records* as having the largest flower structure in the form of an Emirates Airbus 380 (Fig. 2.6).

Across the Persian Gulf, the country of Iran is rather unique in that its gardens are not only some of the most iconic in the world, but also were the model for many Islamic gardens in the Middle East as well as South Asia and even Spain. Nine gardens in Iran are so special that they have been declared World Heritage Sites by UNESCO. The large number of outstanding gardens in Iran speaks to a culture in which flowers are a special part. Thus the Damask Rose Festival in the Kashan region of Isfahan has been cited (Zamani-Farahani and Fox, 2018) as the source of the first rose water tourism event in the world in which large numbers of presumably domestic tourists go to the fields where roses are blooming to pick the roses, collect or manufacture rose water, and sample various products made from the rose water.

Southeast Asia

Southeast Asia runs the spectrum of garden types in the world, from French colonial gardens

Fig. 2.6. Emirates Airbus 380 in flowers, Dubai Miracle Garden, Dubai, UAE, 2015.

in Hanoi surrounding the former French administration buildings and the Jim Thompson House and Garden in Bangkok, to gardens like Nong Nooch in Thailand with elephant shows that are as much theme park as garden.[17] In between there are gardens within zoos (Ho Chi Minh City and Hanoi)[18] and new gardens like Senteurs d'Angkor Botanic Garden in Siem Riep, Cambodia with commercial aspirations for tea and perfume and healing plant sales. Unique to both the region and the world are the gardens of Singapore. Singapore has declared itself the garden capital of the world with the historic Singapore Botanic Gardens, the relatively new but spectacular Gardens by the Bay, and most recently the development in Changi Airport of gardens in Terminal 1 (Cactus Garden), Terminal 2 (Sunflower Garden and Orchid Garden), and Terminal 3 (Butterfly Garden), all of which are connected through the Jewel, a five-story public area with gardens, canopy walks, and restaurants.

South Asia

India

The Botanical Survey of India lists 122 botanical gardens in India. Many are colonial legacies of the British administration but since independence in 1947, gardens continued to be built mainly as research institutions associated with universities. In addition, many areas under the auspices of the forestry department contain significant floral displays and collections that are of great attraction to tourists. Thus, for example, the Valley of Flowers in Nanda Devi National Park, Uttarakhand, in the Gharwal Himalaya, is a World Heritage Site, famous for its wildflowers and which in 2016 attracted over 10,000 visitors in the three months it was open. A trend of opening gardens for tourism, particularly ecotourism activities, has become particularly marked in recent years.

Sri Lanka

There are five botanical gardens in Sri Lanka and much like other Asian gardens, they attract significant numbers of visitors. Peradeniya Royal Botanical Gardens attracts over 2 million visitors per year and Hakgala Botanical Gardens over 500,000 visitors. Henarathgoda Botanical Gardens outside Gampaha has the distinction of having an original Para rubber tree, derived from one of 2700 seedlings brought from Brazil via Kew Gardens in 1878. Much like South Africa, in a plan to provide a botanic garden in all its biomes, a dry land garden, Mirijjawila, and a wet zone garden at Seethawaka were opened in 2014. All these new gardens were situated for maximum tourist accessibility as Sri Lanka, famous as being the site of the original Garden of Eden, uses plants extensively for tourism promotion.

Pakistan

Pakistan has 27 gardens listed as belonging to the Pakistan Botanic Gardens Network. Nineteen are affiliated with schools or universities and the rest are either government research institutes or private gardens (two members). The Secretariat states:

> The biggest challenge that the secretariat is facing today is the insufficient levels of political and financial investment for conducting research in these areas. This is because we have a limited national constituency amongst influential decision makers, therefore we are striving hard to overcome this obstacle by developing productive relationships within and between botanic gardens, conservation groups, Govt. commercial and private sectors, local communities and others interested bodies as valued partners in the conservation and sustainable use of Pakistan's unique environmental botanical and cultural heritage.

Oceania and the Atlantic Region

The gardens of Oceania (and the Atlantic region) are primarily dedicated to tropical and subtropical plants but all have an active program in cultural awareness of the region or environment in which they are located. Thus, for example, the National Tropical Botanical Garden of Hawaii comprises five gardens[19] or preserves dedicated to tropical plants which feature significant educational programs in native Hawaiian plants, history, dance (the Hula), and future sustainability. The quality of the tourist programs at these gardens is such that in 2015 the gardens were awarded the Sunset Travel Award. The island of Oahu is well served by botanic gardens with five gardens that together attract over 470,000 persons per year while the private botanic garden and cultural center at Waimea Bay attracts over 400,000 visitors per year, the majority of which are fully independent tourists/travelers. In other parts of Oceania, gardens have become a major focus of both plant conservation and tourist activity. In Mauritius, the oldest garden in the southern hemisphere and arguably the original source of nutmeg and cloves is the most popular tourist attraction, while in the Seychelles, a major honeymoon destination for Europeans, the coco de mer (*Lodoicea maldivica*) or "love nut" has become a major attraction on the only two islands on which it is found, owing to the prohibition on its export and its linkage to romance and love.[20]

In the Atlantic the two major island groupings in the botanic world are the Azores and the Canary Islands. In the case of the former, they have embraced gardens as one tourism product that they could market to the USA and Europe. The Azores boast four major gardens of which the most important and noteworthy is the Terra Nostra garden that operates as a garden-focused resort exhibiting over 600 varieties of camellia. The new (2017) vehicle for promoting garden tourism in the Azores is the Green Gardens – Azores. The website says:[21]

> The "Green Gardens – Azores" project aims to contribute to the recognition of the historic gardens of the Azores as quality tourism assets associated with nature, culture and wellness, through the discussion and implementation of garden tourism. This project is integrated in the strategic framework "Plan of Action for the Development of Tourism in Portugal 2014–2020".

In the Canary Islands (and the proximate but Portuguese-owned island of Madeira[22]) the plant diversity is legendary, but garden tourism has not been embraced to the degree found in the Azores. Gran Canaria, the most popular tourist destination, has four botanic gardens,[23] while CACT – an acronym for "Center for Arts, Culture, and Tourism" – on the island of Lanzarote has recently (2019) been reaching out to include a nature (botanic) role in the organization, particularly with the introduction into the tourism promotional literature of the Cactus Garden in the village of Guatiza.[24] Its inclusion in an arts and culture brand is a natural extension as it owes its establishment in 1991 to the famous artist Ćesar Manrique.

Notes

[1] Classified by the APGA as Institutional Members.

[2] Which suggests that garden visitation in the USA is growing at over 5% per annum.

[3] Orlando and Vegas vie for the most popular tourist city in the USA with 65–70 million visitors annually. Disneyland and Disney World attract 17 million and 37 million visitors, respectively.

[4] Estimated at 28 million in 2018.

[5] From 2012 through 2020, five separate exhibits that make up Nature Connects®: Art with LEGO® Bricks (the trademarked name) have appeared in 58 public gardens, zoos, and nature centers. All the exhibits will be retired at the end of 2020.

[6] English Heritage's gardens are oriented to the historic nature and relevance of its gardens. Places like Down House (Darwin's home), Kenilworth Castle, and Osborne House (Queen Victoria's summer residence) are presented as essentially heritage properties as opposed to the National Trust's legacy properties. Chapter 10, on gardens and historic homes, will discuss heritage gardens more fully.

[7] In this regard the Insight Department is to be commended. They are one of the few (Fáilte Ireland has also seen the light) tourism research organizations to recognize the value of garden tourism.

[8] As has been noted in earlier publications, while Europe has many gardens (over 800), a significant number are display gardens and do not market themselves as botanic gardens. Thus, there are probably over

400 gardens that call themselves "botanic gardens" and pursue activities common to botanic gardens (research), while the other 400 are gardens in chateaux, Disneyland Paris, and villas which are significant tourist attractions and not strictly botanic gardens.

[9] Unfortunately, this list includes some nurseries, florists, museums, craftsmen, and vineyards. Available at: https://luoghi.italianbotanicalheritage.com/ (accessed July 10, 2020).

[10] Entebbe has particular renown as a site for bird watching.

[11] Hatam National Botanical Garden was established in 2008 and Kwelera National Botanical Garden in 2014.

[12] Somewhat controversially, the NZGT assesses and awards "levels of significance" to its member gardens. Thus, a ranking of six stars would rank a garden as of International Significance. Tourism New Zealand recognizes six New Zealand gardens as having international significance and in 2018 the NZGT added three more gardens to its list of gardens with international significance.

[13] In October 2017, the world's largest botanic garden, Qinling National Botanical Garden in Shanxxi Province, opened after 20 years of planning and development with a total size of 639 km^2. It covers 25 river systems and lies at a height of 480–1500 m above sea level. The Chinese government shut down five hydroelectric plants to provide the garden area.

[14] Xishuangbanna Tropical Botanical Garden is over 50 years old.

[15] The Arab Spring was most marked in Egypt and North Africa starting in 2010 but unrest spread to Bahrain in 2011. While the violence and uprising continue today in North Africa and the Levant, the effects linger in the Persian Gulf, particularly in Yemen.

[16] With upwards of 150 million blooms. There is also a butterfly conservatory on site.

[17] Although the collection of cycads, cacti, and succulents in the garden is one of the most renowned in the world.

[18] The garden enthusiast in Ho Chi Min City's botanical garden is challenged to ignore the awful conditions in which the animals live, unless of course you like to see rats fighting with toothless bears for scraps in the concrete enclosures.

[19] Of the five gardens, three are on Kauai and the fourth on Maui. One of the five gardens, Kampong, is in Coral Gables, Florida.

[20] The reader is encouraged to go to Wikipedia to see why it is called the "love nut".

[21] Available at: https://www.otacores.com/greenga/?lang=en (accessed July 25, 2020).

[22] The island of Madeira with its capital Funchal has five highly rated botanical gardens.

[23] The Jardin Botánico Canario Viera y Clavijo, 7 km from the capital Las Palmas, is the largest (10 ha/27 acres) and most important for its conservation work on native Canary Island flora.

[24] Unfortunately, the Cactus Garden is some 20 miles from Arrecife, the main city on Lanzarote, and hence difficult for independent tourists to visit.

Boomslang or "tree snake" in Kirstenbosch National Botanic Garden, Cape Town. Photo courtesy of Adam Harrower.

Cactus garden, Lanzarote, Canary Islands. Author's own photo.

New welcome centre, Royal Horticultural Society Garden, Wisley. UK. Reproduced by Kind Permission of RHS.

Ocelot caught on camera trap by Panthera in Vallarta Botanic Garden, Mexico. Reproduced by kind permission of Vallarta Botanic Garden and Panthera.org.

Vallarta Botanic Garden, Puerto Vallarta, Mexico. Reproduced by kind permission of Vallarta Botanic Garden.

Jardin Majorelle, Marrakech, Morocco. The Modernist garden of designer Yves St Laurent.

Dwight Grimm, founder of the Projectionists Club Garden at Greenville, New York, presents a lecture and tasting of botanicals to over 100 guests at the US Botanic Garden, Washington, DC, part of the garden's evening outreach programs. Author's own photo.

3

New Directions in Gardens

—————————

© Richard W. Benfield 2021. *New Directions in Garden Tourism* (R. Benfield)
DOI: 10.1079/9781789241761.0003

Changing Tourism Paradigms and Garden Tourism

As a theoretical framework for the charting of new directions in garden tourism it is useful to examine the changing tourism management paradigms that have existed in the past 50 years and the management responses to those external changes. From an appreciation of these evolving scientific revolutions[1] new directions in garden tourism might be suggested.

In the immediate post-war years, and for some 20 years following the war, the prevailing tourism management paradigm was one of creating supply. This was because of a huge demographic explosion of a relatively wealthy young population, the Baby Boomer cohort, following the end of World War II. Thus, theme parks, cruises, increased airline capacity, and lower fares resulted in a leisure explosion that in turn created or permitted new tourism attraction supply. The public garden sector was no different. New gardens in places like Denver, Atlanta, Phoenix,[2] and Los Angeles (Descanso) were welcome additions to tourism supply during increasing populations and a housing boom.

In 1973, and then 10 years later in 1982, recession brought on by oil embargoes (and some say oversupply) created and changed the management paradigm from managing supply to creating demand. Empty hotels and airline seats, dramatic falls in attendance at major attractions, and generally a retrenchment of tourism demand made the pursuit of customers the prevailing paradigm. The garden tourism industry was at this time dedicated to plant collections and while visitation decreased, its impact was usually less than that of commercial tourist attractions.

By the 1990s, economic recovery had led to demand again meeting supply and hence the management need was to differentiate your business from those of your competitors. The way this was to be accomplished was by the implementation of the "customer service" concept in which a business differentiated itself from its competitors by use of superior customer service. Pioneered by Edwards Deming in Japan and first applied by Jan Carlson at SAS in the airline industry, superior customer service became the prevailing paradigm. Some would say this paradigm is still predominant as many tourism and hospitality businesses continue to espouse the customer service concept. However, by the early part of the 21st century, the customer service concept was prevalent throughout the industry and differentiation on the basis of service quality was difficult. It was time for a paradigm shift to create superior business performance.

In 1999 two Harvard business professors, Pine and Gilmore, suggested that what consumers, and especially Millennials, were looking for was not exemplary service but what they called "experiences". Throughout the early part of the new millennium, this paradigm became more and more popular, especially after the recession of 2008 when revenues again fell, demand was muted, and differentiation was sought. In the garden tourism industry, many believed that the spectacular nature of plants, their tactile properties, and their ties to food made them perfect candidates for the experience economy. This paradigm shift, in conjunction with different events at gardens, made gardens more relevant than other tourism products and some say it has contributed greatly to the explosion in garden visitation.[3]

This chapter examines new directions in garden tourism (innovation) by selecting seven major research, product development, and marketing innovations that have characterized gardens in the preceding 7 years. They are:

- gardens and wildlife;
- art and gardens;
- gardens and music;
- Levy walk analysis and gardens;
- plant societies and gardens;
- sensory experiences at gardens; and
- garden branding.

Gardens and Wildlife

A third of the US population aged 16 years or older enjoyed wildlife watching[4] in 2016.[5] People who took an interest in wildlife around their homes numbered 81.1 million, while those who took trips away from their homes to wildlife watch numbered 23.7 million people. Of all the wildlife in the USA, birds attracted the biggest following. There are over 86 million birdwatchers in the USA and 3 million in the UK. Approximately 45.1 million people observed birds around the home and on trips in 2016. A large majority, 86% (38.7 million), observed wild birds around the home, while 36% (16.3 million) took trips away from home to observe wild birds. Participants averaged a startling 96 days of birding in 2016. Away-from-home birders averaged 16 days. While the data do not isolate the number of visitors who come to gardens to specifically (or secondarily, see endnote 5) watch wildlife while at the garden, botanic gardens are included in the statistics as areas of public and private ownership where wildlife watching is a significant activity. Furthermore, gardens in the past few years have proactively embraced their role in the natural world and have increasingly focused on planting and conserving native species for their importance in conserving wildlife species. Missouri Botanical Garden now plants mead's milkweed to provide an egg-laying medium for monarch butterflies and Powell Gardens, Kansas City's Botanical Garden, is growing mead's milkweed to transplant across the state. Much of the initiatives for gardens to embrace wildlife viewing comes from the Certified Wildlife Habitat program of the National Wildlife Federation. Over 50 public gardens in the USA are Certified Wildlife Habitat gardens which, while they number only 50 of 100,000 Certified Wildlife Habitats, provide a significant amount of education on planting, the nature and importance of native habitats, and the plants that belong in such habitats. Because of the importance of gardens as habitat for birds, bird walks are held at many gardens.

One example of a botanic garden that is a Certified Wildlife Habitat and that conducts bird walks is the University of California, Riverside (UCR) Botanic Gardens. Southern California and UCR Botanic Gardens are bird hotspots (see Cornell Lab of Ornithology) because they are on a major flyway. While staff do not notice people obviously birding during a typical day, UCR and the region have many avid and expert birders and bird groups who frequent the garden generally early in the morning when the garden is not busy. The garden offers several early-morning bird walks each year that are guided by local experts, and the site is used for the annual Audubon Christmas Bird Counts. The San Antonio Botanical Garden in Texas is also on a bird migration route/corridor and enjoys significant visitation from birders from throughout the Southwest owing to the fact it has 249 bird species sightings and runs a birding checklist as well as tours and walks for birders.

Many gardens provide bird recognition pamphlets and identification plaques for bird watchers. On the north shore of the island of Oahu, Hawaii, the Waimea Valley[6] provides a free identification guide showcasing 20 birds and the botanic garden on Gran Canaria has identification signs throughout the garden.

In Ecuador, Quito Botanical Garden is also used extensively for birdwatching.[7] The garden is both a site for local birdwatching and a location for migrating birds (September–April). Also an increasing number of visiting birders often come a day before a tour of South America starts and have a relaxed day in Quito for some intro birding, mainly for looking at the resident birds rather than migrants, even though the latter are more interesting for non-Americans. Quito Botanical Garden, like several urban gardens, is typical of a site for ornithology for several reasons:

1. The place is small so while it is not particularly "important" for the number or variety of birds, the usual inter-Andean valley birds are resident there and it is a good migration trap.
2. Local ornithologists estimate there are about a dozen Quito birders and yet over 100 visiting international birders.
3. Because of the importance and number of birds in Ecuador, the Ecuadorian tourist board has produced a brochure for birdwatching self-guided tours and this site is included and important as a starting point.
4. Most local tourism agencies are aware of the botanic garden and its importance and that some nice birds can be seen in the garden, often suggesting this to their clients. In addition, eBird now has hundreds of birds on its checklist for this location, so many birders find the place through eBird.[8]

For mammalian and insect species, the fact that most gardens and arboreta are large and often the only large open green space in an increasingly urban environment makes gardens natural locations for both wildlife habitat and wildlife spotting. Furthermore, many studies show that gardening as an activity for wildlife presence and conservation is increasingly prevalent in garden design and plant choice. The Great British Gardens website has a section devoted to gardens that have a particular attraction as places to see wildlife, while in the USA the National Wildlife Federation produces flyers and website directions for planting plants that attract birds, butterflies, and other wildlife to home gardens, particularly focusing on plants for pollinators, because data have shown a serious decline in availability of plant species that are necessary for pollinators.

In larger gardens, the garden itself is a refuge for larger mammals and often part of a larger ecosystem. At Vallarta Botanical Gardens outside Puerto Vallarta, Mexico, five camera traps have been set up in cooperation with Panthera, to assess the possibility of the garden and surrounding forest preserve[9] being a significant home and corridor[10] for Mexican and Central American jaguars. The cameras have already recorded ocelots and other smaller mammals and night staff have heard and recorded jaguar grunts in the area bordering the garden. The garden is also one of the last remaining locations for the spectacular military macaw and river otters.

Art and Gardens

While art has always had a place in gardens, the last few years have seen an explosion of art in gardens in its many forms. One art form often overlooked are the botanical drawings of plants. The genre originated in the need for the description of exotic plants from faraway places in an era, the 19th and 20th centuries, when exotic plants were the rage yet not readily available for view. The genre not only survived but also thrives in many botanic gardens where botanical art is showcased, in such places as the Shirley Sherwood Gallery at Kew Gardens.

Sculpture is a common addition to gardens and the Mark Dion sculptures of Bartram's expeditions in the southern USA in the late 18th century, exhibited at Bartram's Garden in Philadelphia in 2008, was an excellent example of this. The Hard Rain exhibit, which combined Bob Dylan's music with photographs of 15,000 threatened or extinct plant species, was a major draw for both conservation awareness and marketing in such gardens as Kew, Eden, and

Edinburgh. Finally, blockbuster displays, originating with Chihuly in 2001 in Garfield Park in Chicago and then in over 12 botanic gardens around the world, are the height of special art exhibits in gardens. Each installation was the genesis of significant growth in visitor numbers, in some cases pulling the garden over a million visitors as a result of the exhibition.[11] As this is being written, Chihuly has returned to Kew – 14 years since he first exhibited at the garden. The success of Chihuly has stimulated art installations in many gardens around the world so much so that most gardens include an art exhibit in their annual offerings. Roy Lichtenstein in Fairchild Tropical Botanic Garden in Miami in 2007 and Niki de Saint Phalle's art in Missouri Botanical Garden in 2008 may be an expansion of the Chihuly phenomenon and certainly built on the success of Chihuly. There has been significant diversification in the art installations. These have ranged from the sculpture of Sir Henry Moore (Atlanta Botanical Garden), though light shows (Bruce Munroe at Longwood Gardens), to local artists (Lauritzen Gardens, Omaha) and more whimsical exhibitions (bears at Newfields, Indianapolis).

Nowhere has the integration of art and nature and new directions in visitor development been more apparent than at Newfields in Indianapolis, Indiana (see Chapter 4). Newfields (formerly called Oldfields) is perhaps the best example in the USA to illustrate the marriage of art and garden to stimulate growth in both. The history of the property is instructive as the 100-acre property was formerly part of the Eli Lilley estate. The historic house and grounds were given to the city of Indianapolis in 1966 and in 1970 the grounds become the site for the relocated Indianapolis Museum of Art (IMA). Today the complex contains the IMA, Fairbanks Park, The Garden, the historic Lilly House, and the Elder Greenhouse.

In 2017, to reflect the change from purely an art museum to a historic estate and campus, the governing body embarked upon a branding exercise. As part of the branding exercise and to achieve that goal, they undertook a major study of the Indianapolis leisure market and found some remarkable and persuasive patterns:

•   in examining where leisure dollars were spent in Indianapolis, Newfields in 2017 captured 1% of those dollars but realized they could meet more than 50% of the activities Indianapolis residents pursued;

- 5% of the metro population of Indianapolis was aware of Newfields but only 55% had visited; and
- of their current visitors, 16% of the metro population were making 93% of the visits.

As a brand that reflected the presence of historic houses and landscapes, gardens, performance spaces, and an art and nature park, and home to wetlands, woodlands, and outdoor sculptures, the complex adopted the new name Newfields. Today as a brand, its avowed goal is "To enrich lives through exceptional experiences with art and nature".[12]

In addition, to address the potential market, the data above showed the audiences with a propensity to visit that Newfields was missing. Accordingly, new programs and activities were instituted to bring the potential visitors to the garden. The study also found that when leisure participants sought something new, it was rooted in the familiar such as food, drink, music, and shared interests. As a result, Newfields now features a winter lights event, a craft brewery every Thursday evening, an outdoor theatre, a pop-up teahouse, and art and programs throughout the garden year-round.

Gardens and Music

Much like art, music has become a mainstay in the events held at gardens, often to provide greater earned income for the garden's operation. Chapter 7, on events and festivals, provides further exploration of the music events in Athens, Georgia and Boise, Idaho as well as a case study on the series of outdoor concerts at the Queens Botanical Garden, New York, each of which brings a specific audience to the garden based on targeted marketing of the particular audience desired.

Levy Walk Analysis and Gardens

When executing a search pattern (or any other series of motions that requires travel among multiple locations), an organism has a choice as to the amount of time spent in movement and the amount of time spent still. The organism may, for example, choose to move exactly the

same distance each time it moves, then stay still for exactly the same amount of time. Assuming the direction in which it chooses to move is relatively random, that would result in a search pattern that is called "Brownian motion". Another possibility is to vary the length of the movement (called the "flight") such that a very long flight is followed by a series of short flights, resulting in a pattern called "Levy flight" (Fig. 3.1).

A broad variety of organisms have been found to exhibit Levy flight patterns when foraging for resources (food, shelter, water, etc.). What is particularly interesting is that, when compared to other possible search patterns (like the Brownian motion above, or random waypoint), Levy flight is the most efficient (in terms of cost to resource-finding) search pattern. A great deal of research has been done identifying this pattern in various naturally occurring phenomena, including human behavior. Levy flight has been found in human wayfinding in their everyday walking (traveling to work, grocery shopping, etc.), but to our knowledge no one has ever tracked human wanderings in gardens. If garden wanderings are Levy flight, it would suggest that the experiences in a garden (smelling flowers, birdwatching, admiring vistas, etc.) are as much resources as food, water, and shelter. This of course has major repercussions on things like signage (for education purposes), facility positioning (water stations, food areas), and new garden development planning (where does a garden developer put a new sensory garden or blockbuster show). Thus, while the research has merit in and of itself, it is also very important for garden managers and planners. Only two gardens have released data on garden wanderings. The National Trust analyzed garden walk patterns in order to assess damage to soft terrain and a PhD student at the University of the West of England produced a Levy walk analysis of visitors to the Eden Project in Cornwall. In the case of the former, remedial action in the form of barriers and ropes was used to control the movement, while at Eden, the Levy walk analysis suggested existing paths and destinations (the conservatories) determined that the patterns of movement were not random. In a student project at Queens Botanical Garden, New York, students found that floral splendor and water were the most common attractions for visitor wanderings; what the study also

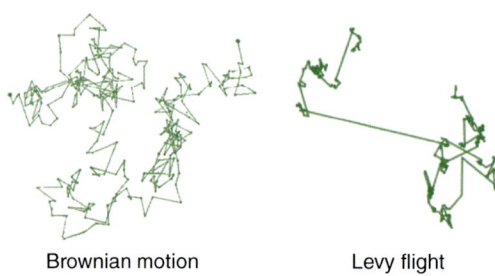

Brownian motion   Levy flight

Fig. 3.1. Brownian motion and Levy flight search pattern.

showed was that signage meant to educate the visitor was often bypassed or ignored in favor of the flower beds or water features.

Plant Societies and Gardens

For over 400 years the science of plants and the study of botany captivated the world. The transportation and display of exotic plants became a major feature of both botanic gardens and their use in home gardens. One legacy of this fascination with exotic plants was the establishment and rise of plant societies, gatherings of people who have a particular love of and dedication to certain plant species. For many years seminars and particularly flower shows attracted a large tourist following. These plant societies still exist today both as national and regional and even local societies. Plant societies in the USA began with the formation of the National Garden Club (NGC) in 1891. Today there are approximately 20 plant societies in the USA – about 50 in the UK[13] – and many have established links with botanic gardens (10 societies), nurseries (5), university gardens (5), private gardens (4), research institutions (1), and plant wholesalers (4). As part of these links the society will often volunteer labor and expertise for the establishment of a reference garden (see Fig. 3.2).[14]

The future of plant societies in the USA and UK is uncertain. Fifty percent report increasing membership while 50% report decreasing membership. In the USA, the Hardy Plant Society of Oregon has 2700 members and runs at least six international tours per year from which it may accrue revenue of US$25,000 per tour. It also runs regional and local tours through a local tour operator. In the UK, the RHS has a dedicated staff member to liaise with and promote dedicated plant societies. However, easy access to the Internet for specific flower or plant advice has threatened the societies' role as a source of plant advice and the adoption of social media for marketing and member solicitation has been slow.

Sensory Experiences at Gardens

In the previous edition of *Garden Tourism*, reference was made to the development of specialized gardens (within gardens) for use by persons with all types of disability and for appreciation by those who were not impaired. These were the so-called "enabling gardens" developed over the previous 10 years to accommodate all types of visitors.[15] Since the development of enabling gardens, many gardens have expanded their *programmatic* offerings to impaired visitors of all types (mobility, sensory, and psychologically) such that they can address their needs as guests, and often specifically their medical and psychological needs (see Chapter 4). Much of these needs are met through sensory experiences and these are described in more detail in Chapters 4 and 6.

Garden Branding[16]

In the realm of branding, the American Marketing Association (AMA) describes a *brand* as "a name, term, design, symbol or any other feature that identifies one seller's goods or services as distinct from those of other sellers". Further, it asserts that *branding* is "a customer experience represented by a collection of images and ideas; often, it refers to a symbol such as a name, logo, slogan, and design scheme".[17] *Destination branding*, including that of a destination attraction such as a botanic garden, takes the notion of brand and branding to a much more unique and

Fig. 3.2. American Conifer Society garden at the State Botanic Garden of Georgia, Athens, Georgia, USA.

granular level. The success of an attraction's branding, at its very core, can be the deciding factor in the decision-making process of a visitor or benefactor to visit or support the garden.

In the book, *The 22 Immutable Laws of Branding*, Al Ries, along with co-author Laura Ries, write that "a successful branding program is based on the concept of singularity. It creates, in the mind of the prospect, the perception that there is no product on the market like your product". Further, as these authors point out, the worlds of corporate and destination branding began to take parallel paths in the late 1990s as it became apparent that there was real economic benefit in the branding of their "product" (Ries and Ries, 1998).

Branding as a concept, one could argue, has been present since the Chauvet cave dweller paintings, some 32,000 years ago. These paintings were some of the first examples of how humans communicated their world into visual meaning. From a marketing tool and programmatic business practice perspective, branding has been present at least since the mid-1800s when the red triangle of Bass Ale became the UK's first official trademark and paved the way for the brand as we know it today. And then in 1900, a first example of a brand as an experi-

ence came about when the tire manufacturers André and Édouard Michelin published the first edition of a free guide for French motorists, along with the early version of the Michelin man branded symbol (Fig. 3.3) (Lippincott Corporation, 2015).

Well-known and iconic companies like Coca-Cola, Nike, and Xerox, in the traditional product world, and Disney and Marriott Hotels, in the services world, have been branding their products, through brand marketing, for decades. For example, Coca-Cola is known as "The Real Thing" while Nike is better known simply by its "swish" icon.

Botanic gardens and branding

It could be said that branding and the focus of the botanic garden, plant life and specimens, go back a very long way; the plant example being tea (*camellia sinensis*).

Twinings Tea Company has used the same logo – capitalized font beneath a lion crest – continuously for 231 years, since 1787. Perhaps even more remarkable, the company has occupied the same location on London's Strand since its founding by Thomas Twining in 1706 (Fig. 3.4). Richard Twining I, Thomas's grandson, commissioned the entranceway, which can still be seen today, and the logo, making it the world's oldest unaltered logo in continuous use according to the company website. Twinings was the first recipient of the Queen's Export Award in 1972 and all bags of Twinings tea carry Her Majesty's Seal.[18]

Tea consumption was not always essential to everyday British life. Coffee, gin, and beer dominated English breakfast drink preferences in the early 18th century. By the turn of the century, however, tea had become extremely popular.

Fig. 3.3. Chauvet cave painting (approx. 32,000 years ago); Bass Pale Ale label (1876); an early version of the Michelin man (1900). From Lippincott Corporation (2015).

After ten generations, family-owned Twinings is now a globally recognized company distributing its tea to more than 100 countries.[18]

There are distinct and differentiating functions of a brand as it relates to the buyer and seller of a product. In their work, "Brand management prognostications", Berthon *et al.* (1999) developed a chart (Fig. 3.5) to convey the message of the functions relative to branding of a product of any goods or service.

Taking this model into the realm of branding of destination attractions, such as botanic gardens, it fits exceptionally well. For the buyer, a successful brand assures product quality, reduction of search, an identifiable symbol or logo around which the brand is built, and the "brand promise" of a memorable experience.

For the seller, a successful brand is a way to provide a coherent message that differentiates the product from its competition. The successful brand should also create brand loyalty and customer return. Specifically, relative to the brand promise, Berthon *et al.* (1999) conclude that it should have the strength to ensure visitor confidence, brand loyalty, and recurring emotional attachment.

There has been work relating to the development of corporate brands and how the resulting brand is different from product brand thinking. Hatch and Schultz explain how the corporate brand contributes to the customer-based images of the organization, not only to those commonly held in product-based branding, but also to images "formed and held by all its stakeholders, including: employees; customers; investors; suppliers; partners; regulators; special interests; and local communities" (Hatch and Schultz, 2003).

Further, to be successful on an ongoing basis, corporate brands should be managed in "relation to the interplay between vision, culture and image". In order to achieve this success, there also must be ongoing and effective dialogue between top management along with

Fig. 3.4. Twinings' trademark and London headquarters.

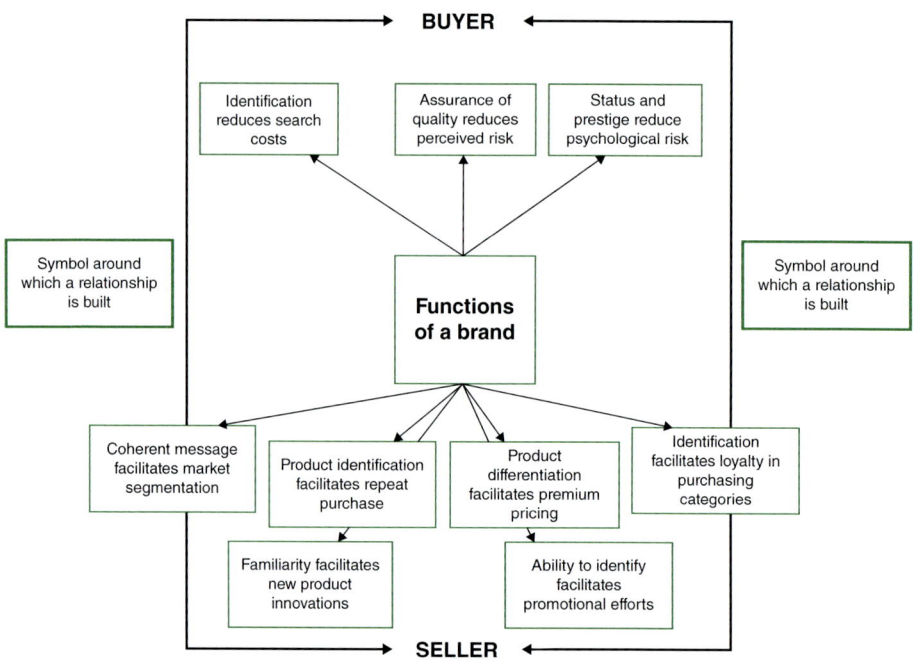

Fig. 3.5. Functions of a brand between buyer and seller. Adapted from Berthon *et al.* (1999).

internal and external stakeholders (Hatch and Schultz, 2003).

On the other end of the spectrum, looking at the relationship of corporate brands and destination brands, Hankinson concludes that key similarities do exist in the management of the two brand categories (Hankinson, 2007). Hankinson further concludes that to be effective, both destination and corporate brands need strong top-level brand management to manage the brand and its sub-brands, along with strong internal and external communications and departmental structures. In addition, both brands must be able to "manage and communicate with a wide range of stakeholders" (Hankinson, 2007).

From this Hankinson proposes a five guiding principles approach to destination brand management, which would closely align with the structures of botanic gardens. The approach first must include strong and visionary brand management leadership, acting as the brand champion, providing a clear vision for the brand, and building and acting upon a strong and clear set of core values, while being able to communicate to all stakeholders. Second, the brand must be, in great part, built by and universally

supported within the organization, from the top of the organization, throughout.

The third guiding principle relates to coordination and process alignment within the prevue of organizational responsibility. It is vital that all areas of the organization are aligned with the brand and in support. This includes boards of directors and other related affiliated groups.

Fourth, very importantly, the brand must be consistently communicated across and through the wide range of outside partner stakeholders. These stakeholder groups include the media and other influencer groups.

Finally, the fifth guiding principle relates to the important act of building, and caring for, strong and compatible partnerships. These partnerships can most often include funding sources and legislative and granting supporters, along with other allied supportive like-minded partners.

This process begins with a solid, strategic vision and the commitment to building and maintaining a brand-oriented culture that permeates throughout the organization and its partners (Hankinson, 2007).

To further understand the importance, complexities, and ramifications of destination branding, this chapter now looks more closely at destination branding relative to the important building blocks of branding: brand image and brand personality. Defined by the educational management study guide, Management Study-Guide.com, *brand image* is "the current view of the customers about a brand and can be defined as a unique bundle of associations within the minds of target customers, signifying what the brand presently stands for".[19] Brand image relies on customer perception, functional and mental connections, and overall impressions of the brand.

Additionally, *brand personality* is defined as "the way a brand speaks and behaves. It means assigning human personality traits/characteristics to a brand so as to achieve differentiation".[20] Brand personality has more to do with "relationships" the consumer may have with the brand and how one might identify with the brand.

In "Destination image and destination personality", Hosany *et al.* (2007) examine the relationship between brand image and brand personality of tourism destinations. Interestingly, they describe their writing and research in terms of the "contentious" relationship between the two concepts.

Brand image and brand personality are both vital to the destination or attraction branding initiative process. Understanding that brand image deals with the functional impressions of the consumer while brand personality relates to the emotional and relational benefits of the brand, these two elements of branding will become essential in the process of research, development, and execution in the brand initiative process (Hosany *et al.*, 2007).

Additionally, relative to brand image and brand personality, Keller (1993) concludes the definition of brand image is "the perceptions about a brand reflected as associations existing in the memory of the consumer" and Batra *et al.* (1993) and Aaker (1997) define brand personality as "the personality traits generally associated with humans that consumers perceive the brand to possess". In the studies done by both, it is readily apparent that both brand image and brand personality are critical to creating destination brand equity. Relative to gardens, both are critical as well.

In light of the requirements for a (garden) brand to be successful, Fiveash (2018) surveyed a small number of botanic gardens to determine the degree branding has played, currently plays, and how much it is or will be a part of their strategic planning, product development, and marketing.

Key common objectives of botanic gardens

In an initial look at the garden sector of the travel and tourism industry, there are five common objectives that emerge, collectively, which a garden organization sets out to achieve. They are:

a. Creating a visitor experience.
b. Education.
c. Conservation.
d. Research
e. Building and retaining a strong donor and patron base.

Fiveash (2018) looked at each, in the context of the efforts of selected gardens' branding initiatives and results, presenting examples from botanic gardens in each of the initiatives.

Visitor experience

Visitor experiences are the greatest external view most of the general population would have of a botanic garden. This visitor experience can be in the form of individual, family, group, or special event visitation and can vary depending on the size and scope of the garden, as well as the location and diversity of the host community.

As in destination branding, visitor experience and the successful implementation of it have a great deal to do with the garden's ability to create an emotional connection or bond with the visitor. This emotional connection has to do with the intent to visit, as well as creating an experience that will capture repeat visitations and word-of-mouth recommendations, to others, to visit the garden (Bethune, 2018).

In his book, *Emotional Branding*, Travis (2000) recounts many successful branding examples that all connect the consumer with emotion-based messages: "You can't make an emotional connection with a customer without using some form of emotion". In addition, there is a tie to the emotional experience and connection to the consumer with the brand promise (Travis, 2000).

The Royal Botanic Gardens at Kew serves as a premier example of this first objective. Kew

Gardens is immersed in creating, implementing, and measuring the ultimate visitor experience for its guests.[21]

Clearly stated in its corporate strategic plan, Kew lays out its vision and commitment:[22] "We aspire to take every visitor on an exciting journey through the diversity of the plant and fungal kingdoms, inspiring them with the importance of plants and fungi in their lives. We connect with visitors through their love for the beauty of plants. We will achieve this through continually improving the gardens, bringing our collections and science to life." Kew even goes several steps further in its visitor commitment, in wanting Kew to be the reason for people to visit the UK and, for British citizens, a reason to travel across the country to visit. Additionally, Kew wants its visitors to be "representative of society and will positively act to ensure there are opportunities for a greater diversity of people".[22]

Education

Education is an objective that botanic gardens take very seriously, across the spectrum of gardens, regardless of size. Gardens offer programs in horticulture and plant science. Most present adults' and children's workshops, classes, and other opportunities to learn about horticulture and nature. Some present more advanced science courses in Plant and Fungal Taxonomy, Diversity and Conservation which include courses that address skills gaps in taxonomy and systematics.

In its 150-year history, the Missouri Botanical Garden, located in St. Louis, has been a great example of the education objective sought by gardens. It has been "committed to connecting people with plants in meaningful ways". Education is paramount to the mission of the Missouri Botanical Garden. It welcomes nearly 1 million visitors per year at its St. Louis area destinations and strives at "strengthening science learning and teaching for students and teachers to helping drive local and global communities towards sustainable living".[23]

Conservation

Conservation is a key objective of every garden, as each wants to foster the sustainable use of plant resources locally and globally. The scope of each garden's conservation efforts depends on the size and reach of the organization and the host community.

From a global perspective, the United Nations World Tourism Organization (UNWTO) defines *sustainable development* as: "Tourism that takes full account of its current and future economic, social and environmental impacts, addressing the needs of visitors, the industry, the environment and host communities". More specifically, the touchpoint at which botanic gardens cross the UNWTO's definition most clearly is in the first of its dimensions to guarantee the long-term sustainability of tourism, namely "Make optimal use of environmental resources that constitute a key element in tourism development, maintaining essential ecological processes and helping to conserve natural heritage and biodiversity".[24]

One of the leaders in conservation within the botanic garden world is the New York Botanical Garden (NYBG), known worldwide as an iconic living museum, a major educational institution, and a renowned plant research and conservation organization. The key conservation efforts of the NYBG are led through its Center for Conservation Strategy (CCS). The CCS's work with local and international collaborators "ensures that science-based policy and decision-making guide and inform conservation initiatives to advance the preservation of life and the sustainability of Earth's ecosystems for the benefit of future Generations".[25] The three main goals of the CCS are to expand, strengthen, and advance conservation initiatives, meeting the mission of the NYBG "to maximize the reach and impact of NYBG's scientific programs and catalyze conservation action that will help save the plants of the world".[26]

Research

Research is a key component to the mission of a garden, no matter the size or reach. Depending on the locale, budget, and mission, the garden's research can provide scientific information, essential to those making key decisions. Because research is tied to not only the future of the home garden but also to the ecosystem it promotes and protects, it is an integral part of the strategic plans and direction of the garden.

A larger garden in the world of botanic gardens, the Missouri Botanical Garden in St. Louis considers itself a world leader in systematics – the study of plants and their evolution. The Missouri Botanical Garden's research "provides scientific information essential to decision makers, from conservation and land use to social and environmental policy". The garden prides itself on its program's accessibility of information and on maintaining the world's largest botanic database.[27]

Smaller gardens like the McIntire Botanical Garden in the city of Charlottesville, Virginia concentrates on research as a part of its strategic plan. The garden's advisory committee is made up of local scientists and allied academics to develop a research plan for the garden to foster and direct research objectives. Further, the garden partners with local and regional academic institutions to fund and direct research programs within the garden.[28]

The Royal Botanic Garden Sydney, Australia has taken a serious and innovative step toward research. The garden has, in its first brand update since its foundation, set out to brand itself to "strategically establish itself as one of Australia's top scientific organizations". The garden sought to brand itself as a place for a scenic and tranquil visit, but one that is grounded in research, and thus communicated a brand "to enrich awareness and make people really think about ways to safeguard their futures" (Thalassinou, 2018).

Building and retaining a strong donor and patron base

The final objective is building and retaining a strong donor and patron base. Most gardens depend on a donor and patron funding base as a consistent funding source to carry out the mission of the organization. Depending on the organization, this can take on a structure which may include corporate sponsors, government funding, and event sponsorships, along with individual donor support. Donor programs can range to programs as large as the NYBG's, which partners with companies "who want to reach our audiences and share our commitment to the environment, education, scientific research, and the arts". The NYBG donor programs range from the US$1500 Patron level to a US$25,000 (and

higher) President's Circle level, with benefits according to donor level. In addition, NYBG has a robust corporate partner program with partners such as Bank of America, Pfizer, Delta Air Lines, and Hyatt Hotels.[29]

Corporate partnerships and foundation gifts are an integral part of the focus of the donor and patron building plan at Lewis Ginter Botanical Garden in Richmond, Virginia. Funding from these sources provides vital operating support for the garden's education programs, Community Kitchen Garden, and world-class plant collections. The garden promotes its corporate partners program as an opportunity to "receive a connection to the Garden, visibility and an experience that makes a lasting impression on clients, employees, and target audiences".[30]

These five objectives, while individually being keys to the structure, development, and viability of a botanic garden, combined, make the garden sector of the tourism industry rather unique in its diversity of mission, with somewhat of a "split personality" that could make the branding of the garden more of a challenge. Relative to this "split personality", Fiveash's (2018) study sought to determine if there were ways to meet the objectives of a multifaceted organization, like a garden, to accomplish successful destination branding.

Branding research in gardens

Using a background in tourism research and destination branding, along with an interest in niche markets of tourism such as botanic gardens, Fiveash (2018) defined three objectives for assessing branding of gardens:

1. Has there been a branding, re-branding, or branding refresh initiative of the garden within the last 5 years?
2. If there has been an initiative, was it influenced by the five areas of focus of gardens outlined above: (a) visitor experience; (b) education; (c) conservation; (d) research; and (e) building and retaining a strong donor and patron base?
3. If there has been an initiative, how important was acceptance/buy-in of the brand by: (a) the local community; (b) the garden's board/governing body; (c) the garden's internal staff

and leadership; and (d) the garden's patron and donor base?

The gardens surveyed

A relatively small but diverse sample of botanic gardens was surveyed, with each being asked to outline the degree branding plays in its strategic planning, development, and marketing of its garden. The gardens were located in the USA, Canada, and Mexico and included gardens ranging from university gardens to government-managed, private, and public/not-for-profit gardens. The gardens ranged in their climatic locations from hot, tropical, and humid to cold and dry. The common denominator, however, among all the gardens, was a passion for "connecting people and plants to improve communities ... to extend beyond traditional boundaries, to have a positive impact and to enrich lives".[31]

Results

1. Has there been a branding, re-branding, or branding refresh initiative of the garden within the last 5 years?

For the purposes of organization, relative to the branding/refresh initiative objective, the botanic gardens were broken out into four categories:

- gardens having no branding or brand;
- gardens having a brand but no brand refresh in the last 5 years;
- gardens having a brand with no brand refresh in the last 5 years, but brand awareness studies done; and
- gardens having a brand and having branding or brand refresh in the last 5 years.

GARDENS HAVING NO BRANDING OR BRAND. Of the gardens surveyed, 23% indicated they do not have a brand for their garden, nor have they initiated any branding or brand refresh in the last 5 years. Of these three, one was a university-funded garden, one a government-funded garden, and one a public/not-for-profit. Additionally, the public/not-for-profit was an international garden.

The fact that many gardens did not respond to the branding study would suggest that this percentage is greatly understated. A cursory content analysis of most gardens' websites would strongly suggest that branding is not part of their strategic planning or outcomes. The

survey showed that funding structure has little to do with the internal or leadership or structural vision needed to move forward with a marketing and development effort such as a branding initiative. However, government-funded gardens may have barriers in their bureaucratic and nontraditional structure that could prevent innovation and vision, unlike an outside, non-governmental structure.

An additional response from one public/not-for-profit respondent was from a director of a botanic garden who said, "We last looked at branding here 12 years ago. Your questions remind me that it's time for a refresh". So, possibly the research helped one more garden realize the value that branding can make to a successful marketing and developmental strategy.

GARDENS HAVING A BRAND BUT NO BRAND REFRESH IN THE LAST 5 YEARS. Fiveash also found that were two gardens in a unique category. Both recognized that the garden did have a brand but had done no branding or brand refresh initiative in the last 5 years, nor had any developmental work, relative to the garden's brand, been done in the last 5 years.

Both gardens have done a significant amount of marketing and development of programming such as seasonal promotions. Each of the gardens has strong mission and vision statements that seem to have been developed with a great deal of forethought in mind for the specific garden's purpose. Both mission statements provide strong elements on which to base and build brand refresh initiatives.

In using terms and phrases such as "elevates quality of life and connects the community through *educational, cultural and social experiences*" (author's italics) in its mission statement, one garden clearly has a good basis to delve into a brand refresh effort, should it choose to do so. The vision statement of the other garden, using terms and phrases such as "*diversity* of plant collections, educational programs and *dedication to conservation*" (author's italics), provides a great start toward a brand refresh as well.

GARDENS HAVING A BRAND WITH NO BRAND REFRESH IN THE LAST 5 YEARS, BUT BRAND AWARENESS STUDIES DONE. Among the respondents, there were gardens which consider themselves as having a brand but have not done any re-branding or brand refresh initiatives in the last 5 years. However, these gardens have done brand awareness studies on the garden's brand effectiveness. One garden, located in a large US metropolitan western area, does a regular semi-annual brand awareness study on which to base any "tweaking of creative/communications strategies". This garden begins with the four core values – "*transformation, relevance, diversity and sustainability*" (author's italics) – which "spell out the Gardens' intentions in the years ahead". These core values are excellent pillars for brand development.

Another garden, located in a subtropical area of the USA, considers that it does have a brand but has not done any re-branding or brand refresh work in the last 5 years. However, this garden is now doing research for a possible re-branding and is in contact with a marketing consulting company to guide the initiative. The garden's mission includes "*diversity by exploring, explaining and conserving the world of tropical plants and … inspiring*" (author's italics) – all bedrock elements to a solid, successful re-branding strategy.

GARDENS HAVING A BRAND AND HAVING BRANDING OR BRAND REFRESH IN THE LAST 5 YEARS. The final group of gardens responding to the survey were a group of very geographically diverse botanic gardens ranging from a non-US tropical garden to a metropolitan area US garden. All but one of the gardens are run as not-for-profit organizations and all of this group has been involved in a branding exercise or brand refresh in the last 5 years. Fifty percent of those gardens used an outside branding, marketing, or consulting agency to help with the branding initiative and 50% undertook the branding in-house. The results for gardens with an existing brand are summarized as follows. The branding was influenced by:

a. Visitor experience – 100% yes.
b. Education – 80% yes, 20% no.
c. Conservation – 20% yes, 80% no.
d. Research – 20% yes, 80% no.
e. Building and retaining a strong donor and patron base – 80% yes, 20% no.

2. If there has been an initiative, was it influenced by the five areas of focus of gardens outlined above: (a) visitor experience; (b) education; (c) conservation; (d) research; and (e) building and retaining a strong donor and patron base?

Each garden was asked to rate the relative importance to the garden's initiative of each the five areas of focus and Table 3.1 indicates this relative importance. It is clear that the major influencers in the decision to embark on branding or a refresh were visitor experience and building and retaining a strong donor base.

3. If there has been an initiative, how important was acceptance/buy-in of the brand by: (a) the local community; (b) the garden's board/governing body; (c) the garden's internal staff and leadership; and (d) the garden's patron and donor base?

It is vital for the owner of the brand – in this case, the botanic garden – to understand that the greatest and most well-thought-out and developed brand ever will never be effective without buy-in. In large part, it is all about the experience of the brand: how will it affect the local community, the board, the staff, the leadership, and the vitally important donor base (Table 3.2).

While referring to destination branding as a concept and subject matter, in "Destination branding: insights and practices from destination management organizations" Blain *et al.* make the case that *visitor experience* and the *brand promise* both play a significant part in the buy-in of the brand. Further, while the visitor may be purchasing the individual products, it is the entire visitor experience that is being sought (Blain *et al.*, 2005). It is that visitor experience that will ultimately drive the botanic garden's brand. The results of the question on buy-in seem to bear this out.

As was referred to earlier but relevant in this context, while looking at the relationship of, and similarities between, corporate brands and destination brands, author Graham Hankinson concludes some interesting correlations to this work on botanic gardens research. Hankinson concludes that for brands to be successful and effective, key management overseeing the brand and its sub-brands must be present and engaged, along with a strong internal and external organizational and departmental structure, able to "manage and communicate with a wide range of stakeholders" (Hankinson, 2007).

Table 3.1. Ranking of importance of garden focus areas to six gardens' initiatives.

Garden focus area	Ranking of relative importance (%)					
	Garden 1	Garden 2	Garden 3	Garden 4	Garden 5	Garden 6
Visitor experience	45	25	40	40	33	90
Education	10	18	12	15	34	0
Conservation	0	20	4	10	0	0
Research	0	19	4	25	0	0
Building a donor base	45	18	40	10	33	10
Total	100	100	100	100	100	100

Table 3.2. Importance of brand buy-in type to six gardens' initiatives.

Buy-in type	Average importance score (range 1–5)
Buy-in by local community	4.83
Buy-in by garden's board/governing body	4.83
Buy-in by garden's internal staff and leadership	4.50
Buy-in by garden's patron and donor base	4.33

Garden branding – conclusions

As was stated earlier, the purpose of Fiveash's (2018) study was to look at botanic gardens in light of the unique and diverse initiatives they represent: a destination product that is a "living" museum, while at the same time being a visitor destination/attraction. In addition, the gardens exist as educational institutions and horticultural research and conservation-focused organizations. The study sought to discover attempts at, and the success of, destination branding in this tourism sector. A relatively small group of diverse botanic gardens was chosen for the survey, to determine the degree branding plays a part in the strategic planning, product development, and marketing in the sector of botanic gardens.

While the sample size was small, it was diverse and represented a fair cross-section of the botanic garden sector as far as size, location, as well as budget and funding model diversification are concerned.

Among other possible observations that can be taken from the research and follow-up conversations, there are three key conclusions:

1. Without exception, the organizational leadership, including the marketing and communications teams, is fiercely committed and loyal to the work and product of the individual botanic garden.

Within the botanic gardens structure, there is evidence of strong commitment to, and genuine fondness for, the work done by the gardens. Beth Monroe, the Director of Public Relations and Marketing at Lewis Ginter Botanical Garden in Richmond, Virginia, is a great example of this type of leadership. Beth and her team have seen that the garden has grown and reached out to engage and attract audiences. She is committed to the garden's mission of education and passion for connecting people and plants to improve communities.

Other gardens have teams of employees and volunteers that are committed to everything from children's education to climate change. Brendan Lange, the Director of Visitor Experience and Marketing for the San Francisco Botanical Garden, expounded upon the garden team's commitment to its work related to climate change. In addition, he discussed new family programs and efforts to reach out to Millennials, led by members of the committed garden team.

2. Organizational/destination branding is not at the forefront of the majority of individual botanic gardens' priorities.

The key takeaway from the study is that organizational/destination branding does not appear to be on the top of the priority list for the majority of botanic gardens. There seem to be common reasons for this: (i) leadership commitment; (ii) financial resources; and (iii) lack of understanding of the value and long-term benefits of branding.

More often than not, lack of financial resources was given as the reason why branding initiatives were not contemplated. Unfortunately, most times there was never the conversation about research being done to complete a cost analysis of branding versus no branding. As one garden said, "the board just somehow knows it will be too expensive".

Further, the volunteer/board leadership of gardens is often unaware of the value of branding and the resulting positive visitor and revenue benefits. The representative of one prestigious garden relayed to me that the board was open to branding but in the most recent strategic planning session, it did not make it to the top ten in priorities.

3. When branding or branding refresh initiatives are undertaken, the two primary influencers are visitor experience and appealing to and retaining a patron and donor base.

The leaderships at the botanic gardens that have done branding or brand refresh initiatives do have a keen understanding that it can and does make a difference in all aspects of the garden's programming, but particularly in visitor experience/attendance and retaining the donor and patron base. Obviously, these are the two most visible elements to a garden's program and revenue-generating portfolio, so these two should be at the top. When taking into consideration the other elements of a garden's programs, education, research, and conservation, all will ultimately be affected by the success of visitor revenue and satisfaction as well as the revenue generated from donors and patrons.

For example, according to Dana Terrazas, Director of Marketing Communications for Desert Botanical Garden in Phoenix, Arizona, "branding and communication" is the fourth pillar of the garden's strategic plan. It was included because of the value and importance the board saw in the branding, visitor research, and programmatic changes, all instituted to "resonate better with visitors". In reviewing the differences

Dana and her team have made in the garden's brand identity, even the most casual observer will see the benefits, both short- and long-term.

The Lewis Ginter Botanical Garden is another example of building a brand with the visitor and stakeholders at the forefront. Beth Monroe, with Lewis Ginter, writes:

> As part of the master site planning process, we assembled internal and external stakeholders to envision what the garden could be for the community. Participants represented a range of diverse groups, including staff and volunteers, community leaders, and even fourth-graders. Some had close relationships to the garden, while others had visited rarely or never. … At the end of the day, we had developed a three-pronged vision to be a garden of: 1) *Timeliness*: A Garden of All Ages; 2) *Community*: A Garden for Cultivating Community; 3) *Awakening*: Mind, Spirit, Body.
>
> (Monroe, 2018)

These are great examples of what can be done when branding is taken seriously, as a means to an end. That end is focusing on the product through the eyes of the consumer, the donor or patron, and the community. Finally, these are examples of what can be done when there is support and commitment of the leadership of the board and key staff members. It can be done!

Recommendations

Based on Fiveash's (2018) study on botanic gardens and the interest, or lack thereof, in branding as a way to increase the marketability, visibility, and recognition of the individual garden, there are two recommendations that could be useful in this area to further the realization of more gardens understanding the potential of successful branding initiatives:

1. Additional research is needed, in an expanded platform, to determine the more broad-based needs, desires, limitations, and institutional marketing predisposition of the individual garden's leadership (board and key staff) toward branding analysis and branding initiatives at the garden.
2. There are very active, knowledgeable, and capable garden trade membership organizations. At present, there does not seem to be any programming within the organizations' purview that includes branding or even marketing

to any real degree. These organizations, both US-based and international, could be the perfect educational venues for marketing and branding discussion and education. These organizations should consider this approach in their future educational programing.

Case Study 3.1: Gin, Tourists, and Botanics in the Experience Economy at Bombay Sapphire

Gin, as a drink, has taken the UK by storm. In 2018 Great Britain consumed 60 million bottles of gin worth £2.2 billion, and growth at over 3% per year suggests revenues will top £3 billion by 2020. The rise has been partially attributable to the popularity of flavored gins and the flavors come from the plants and seeds with which the distillate spirit is infused, the so-called "botanics".[32] Of course, this would be of interest as a new direction for botanic products but the most remarkable growth has been at the distilleries themselves where the time-honored tasting has been supplemented with infusing in the experience economy, with the result that gin tourism is now a recognized subset of tourism.

At the Bombay Sapphire gin distillery in Laverstoke Mill, Hampshire, visitors are taken on a tour of the facility showing the constituents of gin and the historical nature of the distillery.[33] They are assisted with a map containing a microchip capable of reading a voice commentary at the part of the distillery they are currently observing and learning about. But the most significant part is in the Botanical Dry Room where visitors are encouraged to smell the 17 botanics that Bombay Sapphire infuses in the gin, mark on a card their preferences, likes and dislikes, and then, at the end of the tour, they are given a free cocktail containing the botanics they selected as their most preferred. In 2018 the distillery hosted almost 100,000 guests at its facility. Oliver Ward of the Gin Foundry, quoted in *Good Housekeeping* (Chandler, 2017), suggested that gin tourism would be THE major growth factor for gin makers for the foreseeable future, bringing in gin drinkers from around the world.[34] Cheers!

Notes

[1] The use of the term "scientific revolution" is deliberate for it was Thomas S. Kuhn in his seminal 1962 work *The Nature of Scientific Revolutions* that first explained the nature and importance to business and academia of changing paradigms and paradigm shifts.

[2] Desert Botanical Garden in Phoenix was established in 1939 but it was in the post-war years that much of the building and development occurred.

[3] It is clear that the advent of coronavirus has again forced a paradigm shift in garden tourism. What that may look like is examined in Chapter 12.

[4] Wildlife watching is defined as closely observing, feeding, and photographing wildlife, visiting parks and natural areas around the home because of wildlife, and maintaining plantings and natural areas around the home for the benefit of wildlife. These activities are categorized as around the home (within 1 mile of home) or away from home (at least 1 mile from home).

[5] The 2016 Survey counts wildlife watching as a recreational activity in which the primary objective was to watch wildlife, as defined above. Secondary or incidental participation, such as observing wildlife while doing something else, was not included in the survey.

[6] Formerly known as the Waimea Valley Audubon Center and the Waimea Arboretum and Botanical Garden, the Waimea Valley is a historical nature park including botanical gardens. The garden was managed until 2003 by the City and County of Honolulu, when management was assumed by the National Audubon Society. In 2008, management was handed over to Hi'ipaka LLC, a non-profit company created by the Office of Hawaiian Affairs.

[7] The Ecuador Tourism Board has a bird spotting guide that traces a 3-day route throughout the country for bird enthusiasts.

[8] eBird is a web-based resource created by the Cornell Laboratory of Ornithology in which birders can enter their sightings of birds (often as their own "bird list"). From this, locations of bird sightings are mapped,

thus contributing to bird research globally. For tourists who are also birders it provides locations of prime bird habitat, often and invariably, gardens.

[9] The reserve adjoining the formal part of the garden is virtually all forest, currently covering 28 ha from high mountain cloud forests to deep canyons and river valleys. However, at the time of writing, unauthorized dam construction, forest clearing, planning for power facilities, and pipeline rights of way threaten this pristine and protected ecosystem.

[10] See Panthera.org and the proposal for jaguar corridors throughout the Americas.

[11] Denver Botanic Gardens may be the best example, going from 879,982 visitors in 2013 to 1,415,670 visitors in 2014 and making it the most visited botanic garden in the USA in 2014.

[12] And in the process increased earned income by 12% in two years.

[13] There are approximately 50 plant societies in the UK – some of these are one national society that covers the whole country, other larger ones such as the Alpine Garden Society and the British Cactus and Succulent Society have an overarching committee – but very many local groups with individual members. There is no figure for numbers of members, some of the smaller ones only have a few hundred, a few of the larger ones have several thousand – but the RHS estimates there are between 20,000 and 25,000 plant society members in the UK (V. Penn, Wisley, 2019, personal communication).

[14] The most significant reference gardens in the USA are the national herb collection in the National Arboretum in Washington, DC and the national rose garden in Shreveport, Louisiana. In the UK, the garden of David Austin Roses and the RHS Garden Wisley are probably the nearest equivalents.

[15] The need for accommodation of all citizens was a result, in the USA, of the passage of The Americans with Disabilities Act of 1990 that mandated full accommodation in all public areas for Americans with disabilities.

[16] Much of the following is drawn from the first ever study of garden branding by Randall Fiveash in 2018. The author is grateful for permission to use most of the findings.

[17] Available at: https://www.ama.org/the-definition-of-marketing-what-is-marketing/ (accessed July 28, 2020).

[18] Available at: https://www.twinings.co.uk/about-twinings/history-of-twinings (accessed July 28, 2020).

[19] Available at: https://www.managementstudyguide.com/brand-image.htm (accessed July 28, 2020).

[20] Available at: https://www.managementstudyguide.com/brand-personality.htm (accessed July 28, 2020).

[21] See https://www.kew.org/ (accessed July 28, 2020).

[22] See https://www.kew.org/sites/default/files/2019-03/corporate-strategy-0417_0.pdf (accessed July 28, 2020).

[23] See https://www.missouribotanicalgarden.org/media/fact-pages/education-programs.aspx (accessed July 28, 2020).

[24] See https://www.unwto.org/sustainable-development (accessed July 28, 2020).

[25] See https://www.nybg.org/plant-research-and-conservation/center-for-conservation-strategy/ (accessed July 28, 2020).

[26] See https://www.nybg.org/plant-research-and-conservation/center-for-conservation-strategy/about-the-ccs/ (accessed July 28, 2020).

[27] See http://www.missouribotanicalgarden.org/plant-science/plant-science/research.aspx (accessed July 28, 2020).

[28] See https://mcintirebotanicalgarden.org/about/mission-goals/ (accessed July 28, 2020).

[29] See https://www.nybg.org/join-support/corporate-support/current-partners/ (accessed July 28, 2020).

[30] See https://www.lewisginter.org/support/give/corporate-giving/ (accessed July 28, 2020).

[31] See https://www.lewisginter.org/visit/about/mission-goals/ (accessed July 28, 2020).

[32] The botanics are sourced from a variety of regions, but a number of botanic gardens supply botanics to small gin distilleries as left-over cuttings.

[33] The facility was formerly a paper mill making bank notes for the country and British Empire. It only became a distillery in 1986.

[34] The best-selling natural tonic to accompany gin is Fever-Tree Tonic, launched in 2005 and containing quinine sourced from quinine-yielding trees (*Cinchona ledgeriana*) in the Democratic Republic of Congo.

Representatives of the Plants Societies of America tour the National Camellia Garden, Fort Valley, Georgia. Author's own photo.

Garden expansion, Bendigo Botanic Gardens, Victoria, Australia. Author's own photo.

Sampling and choosing botanicals. Bombay Sapphire Gin Distillery, Laverstoke, Hampshire, UK. Photo by the author and used with kind permission of Bombay Sapphire Distillery.

American Conifer Association Reference Garden. State Botanic Garden of Georgia, Athens, Georgia. Author's own photo.

The Botanicals Glasshouses at Bombay Sapphire Distillery, Laverstoke, Hampshire, England. Botanical sampling, selection and tasting are a popular tourist attraction as part of the guided tour of the gin distillery. Photo courtesy of Bombay Sapphire Distillery and used with permission.

4

New Audiences for Gardens

———————————

For many years, gardens (and other tourism attractions) relied on demographic data upon which to describe and address their market, or what might be described as finding out the *who* of garden tourists. Thus age, income, household status, and gender were used because of readily available data from such sources as the census or targeted garden entrance and exit surveys and from which market strategies and garden programs could be developed. In developing new directions in garden tourism, gardens are increasingly using new, more sophisticated, and more detailed market research to delineate *who* their garden appeals to – or, put another way, the nature and motivations of their visitors based on their geographic origins, personal values, attitudes, lifestyles, and behaviors – from which knowledge gardens can target individuals and households who would be interested in their offerings. This emphasis is gradually taking the place of old standard segmentation techniques and thus truly represents a new direction in garden tourism.

Demographic Segmentation for Gardens

Most gardens, based on the kinds of visitor data they collect, seem to have segmented their market into:

1. Primary markets:
   a. An elderly demographic (often referred to as the Baby Boomers, born 1946–1964).
   b. Often or predominantly female.
   c. Higher (disposable) incomes and education.
   d. All age groups with an interest in gardens and gardening.
   e. Usually house owners.
   f. Educational groups (kindergarten to sixth form/K–12).
   g. Tourists, both domestic and international, and day visitors.

2. Secondary or emerging markets:
   a. Millennials (those born between 1981 and 1996).[1]
   b. Those with a propensity to visit gardens as a sedentary leisure-time outdoor activity.
   c. New house owners usually 25 years of age or older.

   d. Generation Xers, born 1965–1980 (a smaller demographic, often with a garden and in need of upgrading landscape).

As of 2017 – the most recent year for which data are available – there were 56 million Millennials in the USA (those aged 21–36 years in 2017) who were working or looking for work. That was more than the 53 million Generation Xers, who accounted for a third of the labor force, and it was well ahead of the 41 million Baby Boomers, who represented a quarter of the total. Millennials surpassed Generation Xers in numbers in 2016. Each demographic cohort is examined in the following.

The Baby Boomers

Although still sizable, the Baby Boomer cohort's sway in the garden tourism world is waning. In the early and mid-1980s, Baby Boomers made up most of the nation's labor force. The youngest Baby Boomer was 53 years old in 2017, while the eldest were older than 70 years. With more Baby Boomers retiring every year and not much immigration to affect their numbers, the size of the Baby Boomer workforce will continue to shrink. Furthermore, the Baby Boomers are downsizing homes, divesting themselves of material goods, and buying fewer goods and services. However, Baby Boomers' impact on travel to gardens will probably not diminish until 2027 when their median age reaches 70 years and thus their number will begin to decrease in absolute terms.

The Generation Xers

Generation X are individuals born between 1965 and 1980. They were born at a time of historically low birth rates in the USA, with the lowest birth rate in 1973. Thus, today they are aged between 40 and 55 years old. They are often described as the "lost" generation and characterized by being disaffected and aimless in their lifestyle. However, they were the generation brought up during the technological revolution in computers and the Internet. Moreover, the fact that there were fewer Gen Xers and they are in demand for a technologically advanced workforce means they are quite wealthy (if they do not have large college debt and a mortgage), and they express that through extensive, often expensive, travel. For gardens they are the highest users of Facebook of any cohort, spending approximately 7 hours on that social media platform each day.

The Millennials

Wake Up Quarterly, a strategic intelligence publication, describes the Millennial generation, those individuals born between 1981 and 1996, as "The most researched yet elusive, demographic group amongst marketers". More than one-in-three American labor force participants (35%) are Millennials, making them the largest generation in the US labor force according to a Pew Research Center analysis of US Census Bureau data (Pew, 2019). The emerging importance of the Millennials is beginning to show in the demographic makeup of garden visitors. In 2017, 30% of the visitors to Missouri Botanical Garden were Millennials and 29% were Baby Boomers, and this pattern is becoming evident throughout the world of gardens.

At the outset, in most surveys probing environmental concern and awareness, Millennials come out stronger than any other group in the areas of overall concern for the environment and specific threats like global warming and its impact on quality of life. More specifically, a majority of Millennials believed they were more concerned than the older generations about protecting the environment. At the same time, though, a majority of adults who comprised those older generations (Baby Boomers and Generation Xers) saw themselves as more environmentally minded now than when they were in their twenties.

As this trend of increasing Millennials into the work force and therefore the tourism industry will increase, Millennials represent a major demographic potential for the future of garden tourism. In a 2016 study of Millennials as part of the Longwood Graduate Program in Public Horticulture, Barton (2017) authored a study on Millennials engaging Millennials within which "engagement" was focused on, or restricted to, Millennials' social media characteristics and habits as they relate to garden visitation.

The study found that 52% of the respondents had visited a public garden and of the 48% of the respondents who had not visited a public garden, 82% were interested. Perhaps not surprisingly, Facebook was their preferred platform and images, videos, and articles were the most desired posting format, suggesting that Instagram and YouTube would gain in popularity in the coming years, but 61% said that social media does not influence their decision to visit a garden. When asked what public gardens could do to encourage more visitation, the Millennials wanted more events (25%), increased advertising to remind them to come (18%), and greater discounts (16%). Strangely, 82% said they wanted to learn more about gardens as an option but only 12% wanted more education (5%) and plant (7%) information.

Of other studies that have focused on Millennials, their attraction to gardening, and their leisure habits, three immediately present themselves. The first was conducted on retail garden sales as part of the National Gardening Survey in 2018. The second was an unpublished study undertaken by Edge Research for the National Trust for Historic Preservation on "Millennials and Historic Preservation" (2017), and the third was a study (2014) entitled "Garden To Table" on food gardening in the USA that specifically examined different demographic groups' participation in growing food.

The annual National Gardening Survey in the USA probes US consumers on their participation and spending on gardening products. In 2017, the older Baby Boomer demographic was still the largest sector of the population who were gardeners, making up 35% of all gardening activities, but in 2018, the growth of the 18- to 34-year-olds that now occupy 29% of all gardening households suggested that Millennials' gardening participation was increasing and significant.

In the National Trust for Historic Preservation survey, when probing US Millennials' attitudes and values, the study found that 97% of respondents felt it important to preserve and conserve buildings, architecture, neighborhoods, and communities, but 62% had yet to be involved in this cause and 53% were interested in getting active. Interestingly, should these Millennials get involved in the cause of historic preservation and conservation, 59% said that food was a key link to history and culture and thus preservation. When traveling, historic properties appeal to almost three-in-four Millennials and 26% make it a point to visit historic places when traveling. Of interest to gardens, almost 60% of Millennials interested in historic properties will go there to visit for a happy hour.

The third study that suggests the increasing and bankable participation of Millennials in gardening was one conducted in 2014 on food gardening across the generations. It showed that between 2008 and 2013 the number of food gardens increased by 4 million and the number of community gardens increased by 2 million. This rise in participation can be attributed almost entirely to Millennials' participation rising by 63% over that 5-year period. As Baldwin (2018) puts it, "Millennials are finally in".

In addition to a changing age demographic for gardens, other demographic changes are beginning to become apparent. In a longitudinal demographic study of visitors to Queens Botanical Garden in Flushing, New York, the garden charted the change in ethnicity of its visitors from 2000 to 2017. The percentage of white or Caucasians rose from 28 to 36%, while the proportion of Chinese, Koreans, and South Asians declined by several percentage points. In contrast, the incidence of black or African American visitors rose from 4 to 11% and visits by those of Hispanic or Latinx background rose from 17 to 21%. Supporting the trend noted above, mean age decreased to 42.1 years, down from 46.7 years in 2000. College-educated visitors also increased (50 to 64%) and household income rose by 30% in real terms (see Chapter 9).

Psychographic Segmentation for Gardens

In April 2010, Facebook launched a platform called Open Graph by which third-party, external developers could acquire large amounts of personal data on Facebook users.[2] In 2013 a company called SLC created an app called "thisisyourdigitallife" in which individuals were paid to take a psychological test based on personal likes and dislikes. The survey was filled out by some 300,000 persons but their Facebook friends were also pulled into the survey, resulting in millions of Facebook profiles. In 2014 the data culled from the psychological surveys and social media[3] were used by a political consulting firm, Cambridge Analytica, ostensibly to influence the 2016 Presidential elections in the USA. While this is certainly not an area of interest in garden tourism, it brought to the general public's awareness the use of a segmentation technique, psychographics, that has a long history in tourism research and is increasingly being used by gardens to segment their current and potential audiences.

In the last 20 years or so, this alternative segmentation method, psychographics, has become increasingly popular in tourism and other business circles, as many believe it is a much better base from which to market a (tourism) product. The method was first called "psychographics" by Demby in 1994[4] and uses values, beliefs, attitudes, and behaviors as the base from which to differentiate visitors. The technique was first used as far back as World War I by differentiating people by looks. For much of the 20th century, demographic segmentation was favored over, or preferred to, psychographics, but psychographics was effectively used to describe tourist segments and their preferred destinations as early as 1974 (Plog, 1974; Plog, 1991). The Stanford Research Institute (SRI) in 1986 in the groundbreaking VALS typology provided further use of the technique, but it was the advent of the computer and its ability to handle large and complex data, especially using cluster analysis, that permitted the technique's wider application. Psychographic segmentation, specifically in garden tourism research, was used in the seminal work by Connell (2004) who indicated motivations to visit gardens were clearly psychographic in nature. She described visitors who came to gardens as "day out visitors" (20%), "general gardeners" (70%), or "special horticultural visitors" (10%). Little in the way of psychographic segmentation was evident until 2012 when the Royal Botanic Gardens, Kew segmented its visitors by way of eight differing psychographic groups (see below).

In 2014, VisitScotland also undertook research to describe its visitors by means of psychographics and in 2017 the National Trust in the UK segmented its visitors into five segments. In the USA, the most recent visitor survey in 2017 using psychographic segmentation was conducted at Newfields (formerly the Indianapolis Museum of Art but containing significant formal and informal gardens; see Case Study 4.1). A detailed examination of the Kew study and the Scottish study, both summarized and published for the first time herein, follows.

Psychographic segmentation at Royal Botanic Gardens, Kew

One of the most recent and promising uses of psychographic segmentation has been undertaken by the Royal Botanic Gardens at Kew in London. It identified eight clusters, all of which, to some degree, have a propensity to visit Kew Gardens. While the study is too exhaustive to detail herein,[5] the characteristics of one segment are repeated below to describe how psychographic segments are characterized and to indicate the kind of visitor the garden might wish to solicit:

> This active, outgoing segment likes to live life to the full. They are always on the lookout for new experiences[6] – unconventional ideas or combinations really appeal to them. Having fun with others is important to them but they want more than a social experience – they enjoy the intellectual and emotional benefits that days out can offer ... they are prolific social networkers.

The Kew study also indicates how a garden tourism marketer might wish to communicate with the eight segments, both in message and channel. In the case of the segment described above, social media would clearly be a major vehicle to market to this group if it were assigned a high priority within the eight segments.

Psychographic segmentation at Royal Botanic Garden Edinburgh

VisitScotland undertook a market segmentation study of specifically the English market in 2014. As VisitScotland states:

> Traditional tools often define consumers purely by demographic or life stage variables that assumes everyone in one age group acts in a

similar way. Segmentation offers the chance to *research ... the wants and needs* of consumers as well as their *specific behaviors*, alongside demographic and life stage information ... VisitScotland's segmentation model is based on a *behavioral approach based on ... motivations, and behaviors, attitudes* toward Scotland ... and their use of media channels.

(Author's italics)

From this research, five tourist segments presented themselves and the RBGE has seized on two segments, named "Engaged Sightseers" and "Natural Advocates", upon which to focus its marketing efforts. The research on engaged sightseers, for example, indicates their level and type of social media participation (lower participation and tend to be "followers" rather than innovators), the major print media they use (magazines and usually gardening magazines), and the challenges and obstacles (signage? distance? information dissemination?) to stimulate greater garden visitation. This research on potential visitors can be compared to survey research undertaken in and by the garden itself, to compare and contrast the current visitation. Thus, in the Edinburgh garden,[7] visitors are predominantly female (60%), over 50% are over 45 years of age, and interestingly 10% identify themselves as being disabled. Again, while the national data showed 42% of English visitors had never been to Scotland, only 20% of RBGE visitors are first-time visitors, suggesting that a new potential market of first-time visitors to Scotland and thus to the garden was available. Finally, it seems that the tourist to the garden spends at least 2 hours in the garden and thus, by extension, after 2 hours goes and visits another tourism location. In the USA, often that "other location" is another garden during their vacation. Thus the fact that RBGE is a founding member of a new national collaborative group called Discover Scottish Gardens – which was set up in 2015 in order to raise the profile of Scotland's gardens and its outstanding horticultural offering, both nationally and internationally – suggests a wise and potentially successful development in Scottish garden marketing because there are about 300 gardens in Scotland which are open to the public and members of a new network will create a collective voice for garden tourism in Scotland.

Psychographic segmentation and historic properties

Psychographics as a method of market research has not been as prevalent in the field of history and historic visiting as in the tourism industry. This may be because, as one historian explained, "people come to historic sites and properties because they are interested in history". While this may be a natural intuitive answer and may obviate the need for further research, the fact is that many visitors to historic properties do not tour the property – rather, they go to the gardens and outdoor portions of the property, presumably to obtain visitation satisfaction. Furthermore, in the study referenced above by Edge Research for the National Trust for Historic Preservation on Millennials' attitudes to historic preservation, many had not visited a historic property but had strong motivations for doing so. Thus, visits to historic properties (with gardens) may not be as intuitive as initially suggested.

Psychographic segmentation at the National Trust

In 2015 the National Trust also segmented its audience in a pamphlet called the Audience Insight Model (AIM). It combined a life stage view of the Trust's audience (primarily by age group and household composition) and what they like to do (mindset and motivation) on their day out. The model, in combining the two, gave the results shown in Table 4.1.

From such a segmentation, the Trust could see what portion of the visitor market is underserved or overserved and tailor its marketing campaign to its most lucrative segments based on life stage and motivation, or what the visitor is seeking.

Geodemographic Segmentation

An area of research that has increasingly become popular as it has shown remarkable results is use of the geographic location of a consumer (obtained primarily through zip codes[8]) to characterize and market offers to individuals and individual neighborhoods. An additional benefit of using geographic location is that it can be combined with census or other data to produce a detailed but descriptive composite picture of an area, often using geographic information system (GIS) technology. Two such geodemographic segmentation methods, both currently being used for garden tourism, follow.

The Claritas method

Most psychographic segmentations also cross-reference their segments to demographic data. A third segmentation method that combines demographic data with geographical location is a product called PRIZM from a company called Claritas.

Claritas was founded in 1961 as a means of segmenting markets by geodemographic data, or as described at one point, "You are where you live". For a number of years Claritas was owned by Nielsen Group, the over 90-year-old company that charts what people watch, listen to, and buy, but was sold in 2017. Claritas's signature product is PRIZM Premier which classifies every US household into one of 68 consumer segments based on the household's purchasing

Table 4.1. The current psychographic market for National Trust garden properties, 2015. Courtesy of the National Trust.

Days out segment name	Percentage of current National Trust visitors	Proportion of UK population (%)	Predominant life stage (age in years)	Percentage of current National Trust members
Home and Family	3	12	Young independents (18–39)	9
Out and About	22	28	Senior plus (over 75)	7
Curious Minds	23	18	Seniors (60–74)	42
Live Life to the Full	24	5	Mature independents (40–59)	23
Young Independents	1	6	Teen family (eldest child a teen)	9
Explorer Family	26	20	Young family (eldest child 0–12)	10
Kids First Family	1	9	Any child (0–19)	9

preferences. PRIZM Premier[9] offers a complete set of ancillary databases and links to third-party data, allowing marketers to use data outside of their own customer files to pinpoint products and services that their best customers are most likely to use, as well as locate their best customers on the ground. PRIZM Premier enables marketers to create a complete portrait of their customers for smarter targeting. PRIZM Premier Segments are numbered according to socioeconomic rank (which considers characteristics such as income, education, occupation, and home value) and are grouped into 11 life stage groups and 14 social groups. Life stage groups are based on age, socioeconomic rank, and the presence of children at home, and social groups are based on urbanicity and socioeconomic rank.

Geofencing

Historically, the application of the research showing a garden's major visitor markets has been translated into the form of paid and unpaid advertising (news and press releases). However, paid advertising has only been the prerogative of larger, more wealthy gardens, while the majority of gardens are smaller, often not-for-profit gardens or governed by a trust and thus without large amounts of money to dedicate to paid advertising. When paid advertising (or garden writers) is sought, placement in gardening and lifestyle magazines seems to provide the best return on investment. This choice is invariably borne out by consumer research. In the case of small gardens, local community advertising may be used, but more often the garden uses a steady diet of press and news releases to impart its information. Furthermore, a major presence in advertising campaigns, often coordinated by the local, regional, or national tourist offices, has been a major source of awareness and marketing. This is now changing with the advent and undoubted staying power of social media (see Chapter 5) and the smartphone.

In 2019, there were 249 million smartphone users in the USA.[10] By 2022, it is estimated that this will have reached 272 million or 73% of the US population.[11] Notwithstanding that the USA is considered the leader in the adoption and use of technology, it is behind the UK, the Netherlands, Sweden, and Germany in smartphone penetration.[12]

The power of these devices has generated massive amounts of new data that can be harnessed by gardens. In an exciting and unique application of the technology, Alexander Babbage,[13] an Atlanta-based location analytics, marketing, and research firm, has used this technology to provide marketing insights, primarily to large-format destinations, districts, and venues such as shopping centers, big box retail stores, museums, and theatres. The research method might very well be applied to gardens in the future.

The research works in the following fashion. Alexander Babbage stores billions of observations of where a consumer physically is, as revealed by their smartphone location.[14] The data are then analyzed for a "GeoFence" based on the location of the consumer (such as "look at all the consumers who were seen at an aquarium"), and then the system will analyze the characteristics of the consumer in that location. The algorithm can determine the home location of the consumer and based on the home location, establish from census data the income, household composition, age, ethnicity, etc. of the consumer. This group can be further segmented by demographic characteristics, where they live, time of day visiting the destination, and the distance they traveled to get there. It should be noted that these data can be supplemented or augmented by onsite research for additional data should the facility require it.

The research has significant advantages. It is specific to a facility or attraction, it requires no equipment or onsite survey work, data are immediate, and it can be conducted in any location or facility no matter what type or shape – including competitive locations.

In essence, the research answers the most fundamental questions for any marketing manager:

- Who are my visitors?
- Where do they live?
- Where else do they go for leisure/shopping, etc.?
- Where do they work?
- How far do they travel to my destination?
- How long do they stay?
- How frequently do they visit?
- Who is visiting my competition?

As an example, Alexander Babbage studied a set of consumers visiting the High Line in New York during 2018 using location-based data gathered from their mobile phones and found their demographic and geographic characteristics based on their home locations. In 2018, over 7 million visitors toured the 23 blocks of the High Line. The results quantified that visitors on the High Line are young, educated, wealthy, and diverse, specifically:

- 16% more likely to be under 34 years old (36% are under 34 years versus 29% in New York and 31% in the USA);
- 55% more likely to have at least a college education (59% have a college degree or higher versus 40% in New York and 38% in the USA);
- 117% more likely to have a household income of US$100,00 or more (50% live in households making US$100,000 or more versus 38% for New York and 23% for the USA);
- probably a demographic reflection of New York City's diversity (51% of visitors are non-white versus 55% in New York and 38% for the USA);
- the average time spent on the High Line is 1 hour and 34 min; and
- visitors come from across the USA, the top five states for US-based visits are New York (69%), New Jersey (6%), California (4%), Florida (3%), and the District of Columbia (2%).

New Audiences in Gardens; Persons with Disabilities

The Americans with Disabilities Act of 1990 ushered in a new era of concern and accommodation for visitors with disabilities to gardens. Much of the initial accommodation was based on physical impairments and accommodations were structural in the form of graded pathways, signage for the blind, raised beds, and the like. Increasingly it has become apparent that there is a need for a wider range of accommodations for individuals with other needs such as autism, memory loss, dementia, and Alzheimer's. Thus in 2019 there were Sensory-Friendly Santa Visits at Daniel Stowe Botanical Garden in Belmont, North Carolina. The garden reserved the hour between 4 and 5 p.m. (when the garden was closed to the public) on Tuesdays during the holiday season for children who needed extra time and attention to visit Santa. During this time, the garden welcomed children with autism and other special needs, along with their siblings, to enjoy a visit with Santa in a compassionate environment that was supportive of their sensory, physical, and developmental needs. Families could also extend their stay to enjoy the garden at dusk, as a Mile of a Million Lights began to illuminate, or they could enjoy the garden for an entire evening. The garden offered discounted weekday "Holidays at the Garden" tickets and a free reservation to visit Santa. In addition, in sensitivity to the visitors' specific needs, families were permitted to take their own photos. Those wishing to have photos taken by a professional photographer were encouraged to schedule an appointment on Mondays.

At the Brooklyn Botanic Garden, early garden opening is provided for autism sufferers, prior to the noise and frenetic activity of the normal opening hours, and memory tours are offered for individuals with memory loss and their caregivers. The garden opens only for these specific groups and is structured as a special guided tour designed for individuals with memory loss and their caregivers. Participants can explore the garden's spectacular displays, enjoy a sensory experience, engage in a hands-on activity, and take home a memento of their visit.

At the national US Botanic Garden in Washington, DC, public programs include specific accommodations for different needs. Their goal has been described as accessibility for everyone and accommodation for all. With a listserv of over 900 members, the garden can reach out to specific needs and audiences. The programs include:

1. Easy access pass, first introduced in 2017 during Season's Greenings, the annual holiday show (late November through early January). During this event, the US Botanic Garden experiences high visitation and the popular model train display requires a separate entry point often resulting in a long and slow-moving line. Individuals and caregivers of visitors who are unable to wait in line due to sensory processing and/or cognitive developmental disabilities may request, in advance, an easy access pass to fast-track through the line. As a courtesy, the garden asks that the party size be limited to four individuals. Additionally, two sensory-friendly programs are scheduled during the holiday show that provide early entrance for registered participants, pre-visit materials, and a take-a-break space.

2. Sensory-friendly programs offered to 30–300 guests (900 on listserv). Started in 2015 and with initial numbers in the 30–40 range, sensory-friendly programs were offered in conjunction with special exhibits. Quarterly early openings provided museum access to individuals with cognitive and sensory processing disabilities and their families by offering early entrance for registered participants, distribution of pre-visit materials, and a take-a-break space.

3. Sensory bags. Sensory bags are available to borrow from the visitor information desk. Many individuals with autism spectrum disorders and/or sensory processing disorders have trouble with everyday sensory stimulation; this can be related to noise, touch, taste, textures, or a variety of sensitivities. Items available to borrow include noise-reducing headphones, weighted lap blankets, and a selection of fidgets to help ease sensory stimulation. These resources are also available for all public programs.

4. Specialized programs. Experienced educators work with both children and adults at all developmental and cognitive levels. A tailored onsite experience is available upon request. Field trips for special education/inclusion classrooms, community, family, and adult day support groups may be arranged and can include enhanced sensory opportunities and tactile experiences. One other type of specialized program are youth programs (13 years of age or above) where youth come into the garden to experience arts and science[15] and develop a sense of living, learning, and inclusion within the community.

5. Routes and reflections programs for individuals with memory loss. This program was in the pilot stage in 2019 with a view to full roll out in 2020. The idea was to use plants along designated routes to spark memory in individuals with memory loss. At the time of writing, in the first program, it was found that eucalyptus smells brought out memory in one person of Mediterranean descent while certain herbs reminded another person, suffering from memory loss, of her childhood years in Cuba. The program will rely heavily on trained volunteers as well as significant interaction and involvement by the carer/companion.

Case Study 4.1: Newfields, Indianapolis, Indiana, USA

In the 1960s one of the world's great art museums, the IMA,[16] moved on to the Oldfield's estate, the former home of E.K. Lilley and his family. The museum was now located on over 100 acres of parkland and contained 53 acres of garden and a 113-year-old historic mansion. Throughout the following 50 years the site became synonymous with the IMA, but in the decade following 2010 the museum was required to grow and increase revenue by examining the competitive landscape to broaden and expand its appeal. The most obvious option and probably with the most potential was the integration of the art experience with the natural landscape around the iconic museum building.

To achieve that end the IMA embarked on a market segmentation study the consultant called a "Jobs-to-be-Won Market Segmentation" (Halverson Group, 2016). The study began by surveying key stakeholders, experts in consumer behavior, and individual consumers who have a proximity to the IMA. Those individuals were local Indianapolis residents, residents of Indiana outside of the Indianapolis metropolitan region, and residents within 180 miles of Indianapolis but not resident in Indiana. The existing membership base was also surveyed. Areas probed were leisure-time activities, motivations, degree of engagement with cultural institutions, perception of the IMA, outdoor experiences, and how they viewed themselves.

The survey found that only 1% of the respondents had been to the IMA as part of their leisure activities[17] but 50% of residents had been to the IMA. Clearly, reasons to visit more often appeared necessary. Also striking was the fact that the 100-acre parkland was relatively unknown. Adults were the primary visitors with a large deficit in children's visits – again an area of potential growth. Finally, and significantly, the study indicated that the provision of a garden would have a great impact on visitation.

The conclusions, based on the empirical findings and the motivational attributes of the respondents, were that there were eight psychographic segments with which the IMA could work to expand its audience:

- "Purposeful Pause" – the segment in which the museum is anchored.
- "Outdoor Escape" and "Mental Reboot" – both segments to which the 100 acres of parkland could appeal.
- The other segments were:
- "Along for the Ride", "Social Celebration", "Pure Fun", "Current and Connected", and "Purposeful Play" – while offering opportunities, these segments were considered less attractive for immediate exploitation.

Essentially, a cohort that wishes to relax and slow down has the most appeal to IMA marketing. Coupled with the patterns of visitation identified in the survey, the IMA has, since 2017, dedicated its marketing efforts to the "Movable Middle", namely visitors who have never visited the IMA, have visited in the past but less often than others, and have an interest in art and gardens. Remarkably when asked what would make these visitors come more often, gardens are the strongest draw. Finally, and as an integral part of the expansion from simply an art museum to a venue that engages art and nature, the name and associated brand were changed to Newfields.

Case Study 4.2: Garden Clubs and Plant Societies

Unlike most other products in the tourism industry, gardens and gardening have very large formal groups or organizations of practitioners and aficionados who avidly embrace gardens and gardening.[18] These people belong to either plant societies or garden clubs that are ubiquitous throughout the USA and the UK. As a group, possibly numbering over 200,000,[19] they represent a dedicated and active segment for garden visiting.

There are approximately 50 plant societies in the UK – some of these are one national

society that covers the whole country, other larger ones such as the Alpine Garden Society and British Cactus and Succulent Society have an overarching committee – but very many local groups. Some plant societies have membership in the hundreds while others may number in the thousands. The RHS estimates there are 20,000–25,000 members of plant societies in the UK.

In the USA there are 51 plant societies and like the UK, some are societies with large active memberships and others are small, specialized, and often somewhat precarious in their future existence. The Hardy Plant Society of Oregon contains over 2700 members and is growing by 8% yearly, while societies like the North American Heather Society with fewer than 100 members is in the process of reorganizing owing to declining membership. Most typical is a society like the American Rose Society with 220 regional member societies and 7100 members.

Much like the RHS in the UK with 25,000 members, the American Horticultural Society is dedicated to the furtherance of gardening and horticulture and has 20,000 members. It provides special admission privileges and discounts at over 320 of the nation's public gardens. More importantly, it acts as the de facto tourism arm of gardeners who wish to travel to gardens both domestically and internationally.

Garden clubs in the USA, defined as those organizations belonging to the NGC, are much more numerous. With eight regions, garden clubs in all 50 states and the National Capital Region, 60 national affiliated organizations within the USA and 330 international affiliated organizations around the globe, and with the 5000 member garden clubs and 165,000 members, the NGC is the largest volunteer organization of its type in the world. As a typical example, the Federated Garden Clubs of Connecticut has 122 member clubs and over 6555 members, with some clubs containing large numbers of members. Branford Garden Club on the Connecticut shoreline boasts over 135 active members.

The influence of plant societies and garden clubs on garden tourism is threefold. First, many plant societies have or have arranged with public gardens to plant a "reference garden" where they can display their plants. The Southeastern Region of the American Conifer Society has a reference garden in the State Botanical Garden of Georgia located in Athens, Georgia; this has the benefit of attracting specifically conifer lovers but is often the focus of volunteer activity. The Herb Society of America has an extensive garden in the US National Arboretum that is used for educational purposes for both adults and K–12 students. Several societies build their own display garden. The American Horticultural Society in Washington, DC has a show garden and the American Rose Society has a show garden in Shreveport, Louisiana. In the UK, the David Austin Rose Garden in Cheshire – while not a plant society garden – attracts over 80,000 persons per year as essentially a focus for rose growers worldwide. The second benefit of plant societies is that they often hold their local and regional shows in gardens. This is primarily a show and recruitment vehicle for the society, but the prospect of a room with spectacular blooms and flower arrangements often stimulates visitors to the garden. The third benefit is, of course, the club or plant society member wishing to visit other gardens. The magnitude of this visitation is unknown but an unofficial survey of a number of garden club members in Connecticut suggests 80% will visit a garden while on extended vacation while 50% will visit a garden as part of, or as a driver for, a weekend or short overnight excursion.

Notes

[1] See https://www.pewresearch.org/fact-tank/2019/01/17/where-millennials-end-and-generation-z-begins/ (accessed August 5, 2020).

[2] In 2014 Facebook developed rules to limit access to these data but the rules were not retroactive.

[3] It has been estimated that Facebook supplied information on over 87 million users.

[4] Significantly it was in a tourist magazine, *Holiday*, that Demby first used the term "psychographics".

[5] The Kew study is still considered proprietary information as it is used to target primary markets for Kew visitors (and of course would have value for competing attractions).

[6] See Case Study 3.1 for the value of experiences as a motivator.

[7] RBGE owns and operates two other satellite gardens.

[8] Zip codes were first introduced in 1963 as an acronym for Zone Improvement Plan – a plan to ostensibly move mail quicker. Refined in 1983 with a 4-digit subset of numbers, today the 42,000 zip codes in the USA are used extensively in marketing to US households.

[9] PRIZM Premier is an update to PRIZM model that featured 66 segments, and it includes new inputs for technology adoption and wealth.

[10] With some 2 billion users worldwide.

[11] By 2020, it is estimated that 93% of US mobile phone users will own and use a smartphone.

[12] In the UK the penetration rate is 82.2%; Australia is tenth with a penetration rate of 68.6%.

[13] The choice of the corporate name is not coincidental. Charles Babbage is considered the father of the computer with his design of a computing machine, the Difference Engine, in 1832.

[14] The apps on the smartphone, with location-based services turned on, leave anonymized "footprints" of where the device is hundreds or thousands of times per day. The system has the ability to eliminate destination workers and "visitors" who stay less than 5 min at a location such as a taxi.

[15] Thus a youth may be able to look for the first time ever through a microscope and learn some of the basics of plant science.

[16] Ironically, as well as a number of paintings by Rembrant, Van Gogh, and Gaugan, the museum is home to Georgia O'Keeffe's iconic *Jimson Weed*, a remarkable 6-ft-high white flower.

[17] While this appears small, the IMA was in the middle of other cultural institutions in appeal.

[18] The American Coaster Enthusiasts (ACE), aficionados for roller coasters at amusement parks, has 5500 members worldwide, for example.

[19] The problem with counting members is that a gardener may belong to both a plant society and a garden club.

Millennials at US Botanic Garden, Washington, DC.

The refurbished temperate house at Kew Gardens is accessible to all guests.

National Herb Garden, National Arboretum, Washington, DC.

Troll Exhibit, Morton Arboretum, Chicago, USA. Photo reproduced by kind permission of the Morton Arboretum.

The Mediterranean Dome at the Eden Project, Bodelva, Cornwall, England. The garden attracts over 1 million visitors annually, making it one of the most popular gardens in the UK. Author's own photo.

5

The New Media Landscape

© Richard W. Benfield 2021. *New Directions in Garden Tourism* (R. Benfield)
DOI: 10.1079/9781789241761.0005

Since the mid-1990s and the boom in the Internet,[1] there has been an explosion of the phenomenon of social media as primarily a part of the mix in marketing of gardens for visitation development. In 2013 social media was in its infancy in garden marketing and little attention was devoted to it as a marketing medium. In 2013 the glossy garden magazine was king in garden marketing; today most gardens would claim that social media is their most important marketing medium and all have a social media staff person either dedicated to posting and researching social media or proficient in its use for their garden. Thus, it seems mandatory to examine what gardens have been doing over the past 7 years and are now doing in the realm of social media, and especially in garden marketing and management for tourist visitation.

Social Media Statistics

In 2020 the Internet had 4.2 billion users out of a world population of 7.7 billion. Of that 4.2 billion, almost 3.4 billion are active on social media. On average, people have 5.54 social media accounts and spend 116 min on social media each day. Facebook adds 500,000 new users every day or six every second. Sixty-eight percent of all Americans are on Facebook and 75% check their page every day while spending, on average, 35 min on Facebook that day. There are 60 million active business pages on Facebook and 5 million active advertisers.

Social Media and Gardens

The first thing that has to be said about social media in general and certainly for gardens is that it is very important to garden tourism, because there is so much "noise" out there that social media allows individual gardens to get noticed. In short, social media levels the playing field a bit – if a garden has great content, it can compete with the likes of top brands to rise above the marketing noise. For example, in 2014, *USA Today* held a contest on the Top 10 Gardens in North America. The Butchart Gardens was the only Canadian venue out of the 20 gardens that were initially nominated. Then *USA Today* invited readers to vote on their favorite garden over a period of a month.

Once that exhortation was finished, the results were:

1. Longwood Gardens.
2. Lewis Ginter Botanical Garden.
3. The Butchart Gardens.

Apparently, Longwood Gardens did very little social media promotion in this contest, but Lewis Ginter, without a lot of marketing budget, was able to race to second place, helped by a very loyal online following both locally and regionally and a decent-sized population center. During the contest, all three gardens were in the running for the top spot. But, in contrast, and during a lag in social media promotion, Butchart was even out of the top ten. None of the gardens put marketing dollars into the campaign – but all were engaged relentlessly on social media and got noticed. According to *USA Today*, it was the second most-voted-upon contest in their history – that's garden tourism and social media in action.[2]

Second, one can say that social media is the first medium available to gardens that can connect with hundreds of thousands of people in real time. Standard print or electronic advertising has never had that ability. Given that close, immediate, personal links are desired by all people in the 21st century and furthermore that a sense of personal values, attitudes, and lifestyles (psychographics) may be the hallmark of defining visitors and their motivation in the 21st century, then social media has the potential to be a major force in garden visitation in the 21st century.

Third, social media lends itself well to garden tourism and vice versa. Why?

1. It doesn't require a huge investment and a lot of gardens don't have a lot of marketing dollars, and those that do have already been earmarked.
2. Gardens are visual and much of social media is all about the visual, which is why Instagram is the fastest-growing social platform and has the highest rate of engagement.
3. Social media is about telling a story (or it should be) through the use of great content – and gardens are replete with stories, deep stories, varied stories – and anyone can use a smartphone

to take a photo or video and edit it right within the phone or the social platform and publish it.

4. With social media, gardens can "brand listen", using free social media tools like Hootsuite and follow hashtags, to "hear" what the public is saying or not saying about their garden – and then join in! One garden is even using social media (obliquely) to decide which photos to put into the calendars that will be sold in the gift shop (an important profit center that was noted in the book, *Garden Tourism*), essentially looking at what people on social media like and then compiling that with other data to make informed decisions.

Having said that, the use and efficacy of social media in gardens has yet to be fully defined, owing to its infancy and a lack of academic studies on this topic. Indeed, this chapter represents the first delineation of social media in gardens known to the author, although the APGA did compile a list and best practices of 12 gardens in North America and Australia using social media in 2015.

The suggestions below come from a sample of British and North American gardens and certainly do not represent any prevailing patterns of use, but it certainly provides an initial assessment of gardens and social media.

Gardens that utilize social media (and a cursory survey of Western gardens shows all gardens have some social media presence, usually a Facebook page) use different social media channels for different reasons.

- The main reason is that each targets a different demographic with respect to garden tourism – who are they after and what social media channels are they using? And it is generally the case that gardens use different social platforms to target different demographics. The so-called Millennials (who will be gardens' future customers) are rabidly using Instagram. Some gardens get more engagement on Instagram than all other social platforms combined! Yet it seems many gardens are late to the game with usually a limited (less than 2000) number of followers – but those that follow, love gardens. One garden went from zero followers on that platform to over 6000 in one year. Facebook is interesting; in one garden, 73% of the garden fans are women but for all of Facebook, women comprise only 46% of Facebook users. Thus 25% of the followers at this same garden are men and 2% have indicated no gender. Furthermore, at this same garden women aged 45–54 years are 18% of followers – but they comprise only 4% of all people on Facebook. Clearly this garden has an audience that is interested in gardens, and that garden particularly. Most interesting though is that these demographics almost exactly match their Facebook equivalent demographics on Sina Weibo (新浪微博) in China!

- Secondary reasons are the type of content (particularly, gardens desire channels that display visuals quite well, for example), immediacy of the information going out, or the activity that is being delivered on the channel (broadcasting "Hey look at me!" or brand listening, etc.).

- Tertiary reasons might be ease of use, ability to schedule, or archival value.

The danger of relying wholly on social media is that the organization does not own the platform; so if the content was deleted (for any reason), the garden would have lost not only the content but all the context around it, often just as important as the content. Obviously, to mitigate this, gardens can regularly archive metrics, take screenshots, and save content – and/or share the same content across multiple platforms.

Following, probably in the order of importance to gardens, is a list of the major social platforms that gardens use. Importance does not necessarily mean that the garden likes or endorses the platform, but sometimes a garden must market in areas it finds individually less than desirable for its product.

Social Media Platforms and Gardens

Facebook

Facebook is the most popular channel on the planet, with 2.3 billion users in 2019, and hence most gardens use it. It shows visuals (photos and videos) incredibly well and has great metrics. It is popular across most age demographics and cultures. It is used in the major tourism markets of Canada, the USA, the UK, Japan, and Australia.

It allows for paid targeting of Facebook users both geographically, demographically and by interest. However, it is a "closed garden" – a term used on the Internet to mean that content is not universally accessible or searchable. It is banned in China (potentially a very large market for outbound tourism in the future) and its organic reach (reach without paying) is declining (1.2% for brands, which means that only 1.2% of posted content ever gets to people that are following the Facebook page). Facebook is now a public company with shareholders and so gardens now have to pay, but for smaller entities this just may not be effective, feasible, or cost-effective.

Most gardens use Facebook to broadcast events, pretty photos, and videos – mostly to great acclaim. In turn gardens get valuable feedback in terms of likes, shares, and comments from around the world in real time. Similarly, the garden can engage followers in real time and in their own language (translation function embedded). A garden can test ideas and photos for traditional marketing on Facebook to get feedback and ask those most interested in the garden experience what they think about ideas or plans that are being considered – all valuable information.

What is also of great benefit is one can immediately see what is popular – and invariably what people want to see are lovely photos of flowers in gardens. However, most gardens go beyond that and craft the message with informational posts and posts about non-flower/garden points of interest (such as food or the gift shop). This is why gardens also use other platforms.

As an example of Facebook's power, The Butchart Gardens in Victoria, British Columbia, Canada created a post in 2013 that unexpectedly went viral. Initially people, and then radio and television stations, were asking for photos of a rare snowfall that had fallen on December 5, 2013 and the garden promised them they would get back to them once a photographer had been out to take pictures. Butchart subsequently posted a photo of snow in the garden. The reach on this single photo was 75,000 people, almost 7000 likes, comments, and shares, and 6500 actual post clicks. It is apparent that in rare instances, a fledgling brand, such as a garden, can really rise above the noise and chatter on social media. To give perspective, the largest local paper in Victoria, British Columbia has a circulation of 58,000. Myriad conversations came out of this single post, keeping The Butchart Gardens top of mind for the day and the weeks after.

The Butchart Gardens uses hashtags[3] extensively, mostly to index posts and to get noticed by people outside of their social media channels. They use three "branded" tags: #butchartgardens, #butchartchristmas, and #butcharttraffic (to give real-time traffic updates to people waiting in line for special events and on high-volume days), among many others. This gets like-minded individuals into a conversation together, so in the gardening world, #gardens, #gardening, and #gardenchat are very popular hashtags. Use of hashtags also permits the aggregation of tourist entities to attract followers to a "like clause". For example, the city of Richmond uses the hashtag #RVA, thus Lewis Ginter Botanical Garden can join in a city-wide social media presence by using that hashtag.

Finally, it should be noted that Facebook and other social media have the ability to share content with other tourist and business organizations in a collaborative fashion for post exposure to a wider potential market. Thus, chambers of commerce and destination marketing/management organizations (DMOs) become natural partners using social media. The image or post may tell people nothing about the specific garden; however, garden visitors usually stay for less than 2 hours in the garden and so it tells everything about the region the garden is in – people interesting in traveling to the region find that sort of ancillary, but pertinent, information important. Social media is very much a reciprocal business – it is formally called "building social capital" in the social media world – and therefore important.

It should be noted that many Facebook users gauge success on the number of likes or followers, but that does not equate to return on investment. It is engaged followers that counts. Again, The Butchart Gardens noted in one week that those who were actually engaging in terms of liking, commenting, and sharing were the older demographic – in fact, older and more female on average than their fan base suggests. Furthermore, Japan pops up in the top five user groups – these are mostly Japanese expats living in Victoria.

Most important, for most gardens, is that not only can they reply to people's comments, they can do so at any time and they are notified – this starts a conversation. Others might join in too. The garden can add or correct information or simply say thank you. People like this, especially from brands, and it makes them much more likely to have that brand at the top of mind when considering gardens as a tourism/leisure option.

In summary, the average Facebook post has a lifespan of 3 hours before the Internet forgets about it – it is still there, but it seems few gardens are engaging with it as cleverly as they might.[4]

Twitter

There are 326 million Twitter users in the world with 47 million in the USA. For gardens, Twitter is mainly used for broadcasting important information regarding an event, a change in hours, to engage followers, to brand monitor, or to get other information usually in a short time frame by asking followers or other Twitter users. This is why many gardens have called Twitter "the urgent email in the web". In a typical tweet a garden might have pinned its hours or told visitors on their way to the garden that the parking lot is full! The next tweet may engage followers with a cute photo taken in the gift shop (this lets people know there is one without overtly selling; it should be noted that using social media to sell something specific is frowned upon – better to market and give people information and options – so one won't see a garden posting prices or saying things like: "Buy now at $25.50!"). The third tweet might promote the transportation option of a partner that brings people to the garden, especially if that parking lot above is full. It gives a link for more information and is hashtagged. A fourth tweet is a call to action to look at someone else's photos, engaging them by "retweeting" their original post that started out on Instagram, was posted to Twitter where the garden was tagged, noticed it, and is in effect saying: "Look at this, someone else besides us thinks this is at least worth looking at, if not visiting!".

The average tweet has a lifespan of a few minutes unless it gets retweeted a lot or unless it is pinned (only one allowed) to the top of the page. One can breathe new life into a tweet by retweeting it – copying and tweeting – but you never do that to your own tweets, that's like

drinking the contents of your finger bowl at a state dinner. Engagement on Twitter for gardens is seasonally specific, so in a higher season such as fall there is much more engagement and value than at the beginning of December. Initially Twitter was not necessarily visual, but now through links one can send photos and videos to be viewed. These have to be clicked for the most part, although photos are now starting to show up without active engagement by Twitter users.

Instagram

Instagram has over 1 billion users and is one of the fastest-growing social media platforms, especially for gardens. Instagram, a photo-sharing app, is highly visual and, not surprisingly, gardens – highly visual places – are seeing great engagement with Instagram. It is almost devoid of advertisements. It gives few demographic statistics, but it is HEAVILY weighted in the 24–34 years age range. These are the Millennials, the people who are stereotyped as wanting everything yesterday and are not afraid to say so, and they are mobile (Instagram only works on a smartphone or tablet[5]). They want choice and they want it their way.[6] Almost all gardens yearn for Millennials but they, like McDonald's, are facing the problem of attracting them. Many gardens are trying to start the engagement process with them now because they are the garden's future customers, and because they love engaging and they love to love brands and aren't afraid to tell the world about it – if you are doing it right. If you aren't, watch out – it's potentially a two-edged sword. There are no gardens known for paying for followers/fans – never do that! For one garden, of the top five photos posted on Instagram, four were landscapes and only one a close-up shot. Interestingly, only one is of flowers. An additional benefit of Instagram is that Iconosquare and other free metrics tools may be used for measurement tracking. Contests on Instagram have been extremely effective both as a means of attracting visitors as well as a means of assessing success. More practically, contests allow the buzz on the garden to be extended and the garden can also use the photographs in its own marketing. The use of hashtags # with keywords is also an important element of Instagram.

The average Instagram post has a lifespan of several hours, mostly because of hashtag searches which are popular on the platform. But a garden often will put in more hashtags later in the comments section which adds new life to an old Instagram post.

Garden blogs

A blog (a truncation of the expression "weblog") is a discussion or informational site published on the World Wide Web and consisting of discrete entries ("posts") typically displayed in reverse chronological order (the most recent post appears first). Until 2009 blogs were usually the work of a single individual, occasionally of a small group, and often covered a single subject. More recently "multi-author blogs" (MABs) have developed, with posts written by large numbers of authors and professionally edited. MABs from individual gardens account for an increasing quantity of blog traffic. The rise of Twitter and other "microblogging" systems helps integrate MABs and single-author blogs into societal news streams.

The majority is interactive, allowing visitors to leave comments and even message each other via graphical user interface (GUI) widgets on the blogs, and it is this interactivity that distinguishes them from other static websites. In that sense, blogging can be seen as a form of social networking service. Indeed, bloggers not only produce content to post on their blogs, but also build social relations with their readers and other bloggers.

At the time of writing, the top gardening blogs in North America according to BlogRank[7] are:

1. Backyard Gardening Blog.
2. GardenRant – primarily written by Elizabeth Licata, Amy Steward, and Susan Brown,[8] and covering not only the art of gardening but also garden tourism in general.
3. North Coast Gardening.
4. Gardening Gone Wild.
5. Busy-at-Home.
6. Ramblings from a Desert Garden.
7. Veggie Gardening Tips.
8. Shirley Bovshows Eden Makers Blog.
9. Cold Climate Gardening.
10. Growing with Plants.

In the UK, the top blogs[9] at the time of writing according to Vuelio[10] are:

1. The Blackberry Garden.
2. The Middle-Sized Garden.
3. Growing Family.
4. Mr Plant Geek.
5. Vertical Veg.
6. Two Thirsty Gardeners.
7. The Frustrated Gardener.
8. The Enduring Gardener.
9. Garden Ninja.
10. MandyCanUDigIt

GardenRant averages 80,000 page views per month and has 11,000 dedicated followers. The site also has a profile of its viewers[11] but unfortunately does not indicate how many of the readers are driven to gardens as tourists by the blog. Nevertheless, Ms. Licata indicates that because she has such a loyal following, she knows most go to gardens that have been featured or mentioned in GardenRant.[12]

Google+

Google+ was closed by Google on April 2, 2019, but prior to closure it allowed users to back up their data. Google+ was used by gardens primarily for archival purposes and to get Google to notice that a garden had made an important blog post, so it was useful for search engine optimization. Some gardens may have had fans on Google+ but for the most part it was believed they did not, so Google+ was often only used in this limited format. Generally, it worked such that when one posted to Google+ from a garden blog, Google indexed this immediately, associated it with hashtags, and anyone following those hashtags would notice that a post or entry had been made on the blog. A garden's blog was key to getting noticed by Google, because Google was looking for good new content. At the beginning of 2016 one garden got 200 views per day on its blog; when Google+ closed the garden was getting upwards of 1500 views of its blog daily. Because it was informationally dense, this was important. Gardens would post photos there too and stream these to social media, which then enticed the potential consumer to their website – one of the pillars of social media marketing. It is true of all social media that one can have a great website with great content but if no one knows about it then the purpose of having such a site is defeated. Students are taught to think of social media as waving a red flag in front of a bull – it works! For example, outside of organic search, Facebook is the second most important driver to The Butchart Gardens' website after its local DMO, Tourism Victoria, and most believe that with the demise of Google+, users and gardens gravitated to Facebook as their social medium of choice.[13] Since the closure, Google My Business has become an important consideration for gardens, but its importance and efficacy are as yet unproven (see Case Study 5.3).

YouTube

YouTube is considered by many as the social medium of choice in the immediate future. Some 300 hours of video are uploaded to You-Tube every minute and the average person watches 40 min of YouTube video content every day. Ninety-four percent of Americans aged 18–24 years use YouTube, and more than half of YouTube videos are watched on mobile devices. YouTube is something that gardens have not done a lot with, although one North American garden has almost half a million views. It is not, as yet, a key component of social media marketing but it is considered to be an important platform for the future because gardens and garden activities are uniquely suited to video, whether it be a 6 second clip or a 15 min video. To date several gardens have posted videos to YouTube and into their site to showcase new developments, seasonal attractions, and even fundraising, but large-scale growth and usage have not yet occurred.[14] While one of the benefits of video is that an individual's mobile device has video capability, most video is shot by traditional media that is streamed on to a garden's site or over social networks as a conversation starter. If using traditional methods, there is the possibility of adding advertising or some other direct message into the video, which makes it very important for getting out a dedicated message. Demographics vary widely but they are very subject dependent.

Pinterest

Pinterest is new for most gardens. In the past it used to highlight important branded accounts or regionally significant accounts. Lewis Ginter Botanical Garden in Richmond, Virginia does Pinterest well – it has over 2.5 million followers, partly due to the way Pinterest is used to highlight accounts. As it is highly visual with a predominantly female audience,[15] Pinterest is a great fit for a garden. Gardens use it to drive web traffic while highlighting beautiful blooms, nature, weddings, travel, and more, and are careful to feature other gardens and sites as well. Balanced content, making sure it's not always "all about the garden", is important in an overall social media strategy.

The average lifespan of a pin is less than 3 min, but they are working on search to make pins more relevant and searchable, and Pinterest will probably remain a key focus for many gardens in the coming years.

Tripadvisor

Tripadvisor is a consumer review site, not a social medium, and while not a social media site that the garden can control or post to,[16] it is integral to the marketing of many gardens and should be for any garden. Essentially Tripadvisor is used as a kind of third-party confirmation of social authority/license: "Look here, see what people are saying about us – organically – without our involvement, and by the way, they LOVE us!". For example, Bombay Sapphire, the gin distiller in the UK, with the botanicals tasting room, cites a favorable Tripadvisor rating of 4.6 out of 5 for the experience at the distillery, as confirmation of customer satisfaction.

Ranking Social Media

In an informal research project, several of the gardens most heavily using social media were asked: "If you had to give up all your social media except one, what would you keep?" Instagram appeared to be the most popular social medium, and most desired to keep Tripadvisor as it was not a social medium. One commented that if pressed on that point they would take Tripadvisor over five Facebooks.

Case Study 5.1: The Media Mix at the Eden Project Integrating Social Media and Print Media

As was noted in Chapter 1, the Eden Project in Bodelva, Cornwall was desirous of exploiting a new marketing niche group of persons interested in horticulture. The current market was 162,000 strong but with 27 million gardeners in the UK, Eden was of the opinion that this represented a significant growth potential. Strategically the garden had to position itself anew in the minds of the market segment. Thus, the marketing should reflect:

- What? A richly textured contemporary garden.
- How? By inspiring and exciting.
- Who? Gardeners who want new ideas.
- Where? In a striking setting.
- Why? To appreciate imaginative horticulture.
- When? Throughout the seasons but especially in spring (March–June) and fall (August–October).

Marketing methods: print, digital, and in-house

Two major print publications, *Gardeners' World* and *Gardens Illustrated*, were used for full-page, color print advertisements reaching 250,000 persons and the message was strongly oriented to horticultural imagery, promotions for a "Two for One" entry, and individual messages of horticultural interest (orchid displays, new garden openings, etc.).

In the digital marketing campaign, new creative content was sent out by solus email[17] offering competitions and other promotional pieces. On Facebook, paid advertisements were targeted to Cornwall and Devon garden-interested users using garden-related words. For the organic reach campaign, 45,000 impressions were registered while for the paid search, 7000 impressions per burst of activity were recorded. Finally, Google Display Network was used on gardening sites and over 181,000 impressions were recorded.

The in-house seasonal campaign and events were placed on the Eden website and reinforced by daily posts on Twitter and Instagram.

This was coupled with email blasts from an Eden-owned database of 35,000 emails and videos of garden developments and news.

The result, that almost certainly contributed to the rise in visitation of over 2% per annum, was a rise of over 50% in visits to the horticultural portion of the website, over 5000 redemptions of the two garden entries for one offer, over 7000 entries for the competitions, and significant rises in numbers for all events.

Finally, in a marketing campaign that makes use of the power of social media to get people talking, the Eden Project in June 2016 "produced and marketed" a new men's perfume, Titan, based on the olfactory characteristics of the flower titan arum (*Amorphophallus titanium*), nine of which were blooming at the time of its launch. The question of whether it was a joke[18] consumed the Internet for a long time ... was it a joke?[19]

Case Study 5.2: Ho'omaluhia Botanical Garden, Hawaii becomes an Instagram hit

Ho'omaluhia Botanical Garden is a 400-acre garden outside Kaneohe developed by the US Corps of Engineers as a flood protection area owing to the high incidence of rainfall on the north shore of the Island of Oahu. The stunning Ko'olau Mountains provide a spectacular backdrop to the garden but following its opening in 1993 it received a limited number of tourists, usually in the 100,000 range or around 10,000 per month. That all changed in 2017. In early 2017 the garden was a PokéStop with Gyms for Pokémon GO[20] players. Attendance figures initially reflected this craze but with the announcement in the fall of 2017 that the Ko'olau Mountains backdrop to the garden was one of the top ten locations for Instagram selfies, attendance rose from 215,573 to 323,184 or a growth of 50% in three years (Fig. 5.1).

Case Study 5.3: Lewis Ginter Botanical Garden and Google My Business

As noted above, Google+ was a media platform up until 2019 and although engagement there was low, gardens often used it to help with search

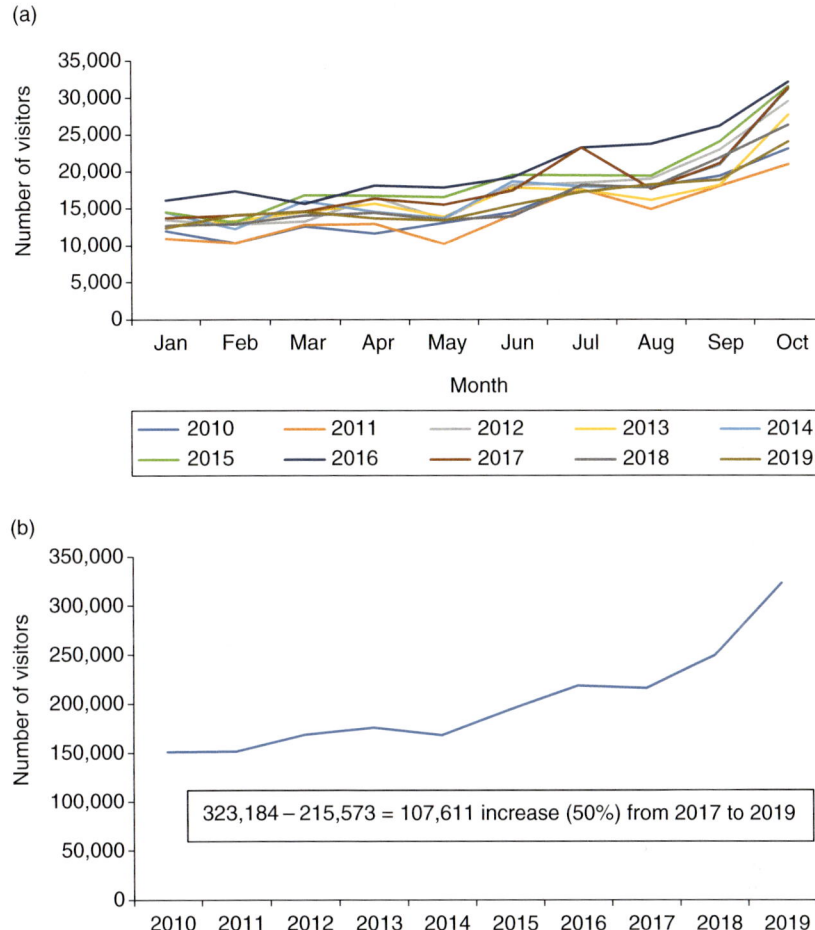

Fig. 5.1. Annual (a) and monthly (b) visitor attendance at Ho'omaluhia Botanical Garden, Oahu, Hawaii, 2012–2019. Data courtesy of Department of Parks and Recreation, City and County of Honolulu, Hawaii.

engine optimization for their website and blog. Now gardens are gravitating to Google My Business instead. Although Google My Business has been around for quite a while, the ability to add posts, "offers", events, and even sell merchandise on Google My Business is relatively new. In addition to helping with search engine optimization, posts on a garden's Google My Business page show up on the knowledge panel and Google Maps. If potential customers Google a garden, these updates, along with visitor reviews, are one of the first things that they will see. Since garden visitors typically check hours and directions on Google

before they visit, having the ability to quickly post an update in this highly visible area is extremely helpful. During the Covid-19 pandemic Google My Business added the ability for gardens and other businesses to "temporarily close", to inform their guests quickly during a time when things were changing rapidly from day to day.

This, of course, creates a one-stop shop for the garden and enables the garden to post into Google information it wants the potential customer to have without the bother of searching the garden's often lengthy website. In the case of Lewis Ginter Botanical Garden, which (as has been seen) has been using social media effectively for the last decade, it is now a key platform to post content and read and respond to visitor reviews.

Google My Business integrates with Google Analytics (website traffic) and Google AdWords (hopefully you've gotten the Google for Non-profits US$10,000 per month Google Ad Grant) and allows you to manage multiple locations from one dashboard (i.e. the garden shop, café, or secondary location) and view insights so you can evaluate your effectiveness.

Google My Business info is crowd sourced. It can come from third-party platforms (like Wikipedia), visitors, or truly anyone who makes an edit. If you manage your Google My Business profile, then you have the opportunity to approve and manage changes. That alone is enough to keep tabs on your profile, but once you create a profile you will have an influx of reviews to respond to, and even text messages to reply to about your business. The life of a Google My Business post is about a week, and if you don't post each week you get a reminder to do so.

Lewis Ginter Botanical Garden considers Google My Business a key platform. To understand the reach, here is an example from early spring 2020. On any given day in spring Lewis Ginter Botanical Garden appeared 1260 times on Google search listings and 17,546 times on Google Maps. On a given day in May, even while closed for the Covid-19 pandemic, the garden saw 146 web visits, 35 people requesting directions, three calls, and one text message from Google My Business. Yearly, while open, that translates into roughly 25,000 website visits from Google and 17,880 requests for directions, along with over 2000 calls and over 100 texts. Gardens can view what words visitors are searching for that help them find their business, and even where the directions inquiries are coming from. This information can inform other marketing and advertising campaigns. That number compares to 35,440 website visits annually from Facebook including paid campaigns, averaging fewer than 100 click-throughs per day on Facebook.

Two of the most significant advantages of Google My Business are first that it provides an immediate response to the most frequently asked questions (Are they open? Directions? Is there a fee to enter?), and second it provides a snapshot picture of the garden with visitor-submitted photos and reviews of the garden comparable to Tripadvisor.[21]

Case Study 5.4: Summer Rayne Oakes and the YouTube Generation

Fresh out of Cornell University as an Environmental Studies and Entomology major and becoming resident in New York as the world's first Eco Model, Summer Rayne Oakes started a YouTube series called *Plant One on Me* in February 2017 as almost certainly the only Millennial talking about houseplants and their care. In 2018 her unique offering went viral with a 60% increase in followers, most of whom were probably were attracted to the search engine for houseplants. By 2020, her 900 YouTube videos were getting 2 million views per month, she had 250,000 subscribers, and 500,000 followers; and as the recipient of two Webby Awards – to date she has received two nominations for a Webby Award[22] and has been a winner twice, in 2019 and 2020[23] – she may well be the most successful garden-focused YouTube-dedicated person in the world.[24] She attributes her success and popularity to the fact that she provides the basic information delivered as basic education with an element of entertainment. She also provides information on plants as part of larger conservation issues as well as human interest stories. Her output is prodigious. She produced a 365-day series on houseplants for the beginner gardener to learn about individual plants, but it is in the area of tours and field trips that Summer is so influential in the use of YouTube for garden tourism. She has produced long-form[25] videos (released two per week) on such garden destinations as Singapore, the Netherlands, Indonesia, Thailand, and domestic gardens in Florida, Missouri, and Texas. For the field trip video on cycads at Nong Nooch Tropical Garden in Thailand she got over 26,000 views and for the series on hoyas also set at Nong Nooch Tropical Garden in Thailand, over 36,000 views. She has also featured plant nurseries, individual botanic gardens, and natural areas as places to visit.

Notes

[1] The sharing of information began as early as the 1970s (with BBS) and CompuServe in the 1980s. AOL first began the widespread sharing of personal information, but it was Friendster (2002), Myspace (2003), and finally Facebook (2004–2006) that heralded the boom in social media as a marketing tool.

[2] In 2019, *USA Today* ran the same type of competition but for the Ten Best *Botanic* Gardens in America. The results were:

1. Minnesota Landscape Arboretum, Chaska, Minnesota.
2. Missouri Botanical Garden, St. Louis, Missouri.
3. Bok Tower Gardens, Lake Wales, Florida.
4. Lewis Ginter Botanical Garden, Richmond, Virginia.
5. Vallarta Botanical Gardens, Puerto Vallarta, Mexico.
6. Cheekwood Estate & Gardens, Nashville, Tennessee.
7. Atlanta Botanical Garden, Atlanta, Georgia.
8. Chicago Botanic Garden, Chicago, Illinois.
9. San Diego Botanic Garden, Encinitas, California.
10. Montreal Botanical Garden, Montreal, Quebec.

The gardens enjoy the recognition but think that the ranking does more for their image with residents of the city than stimulate visitation. It is believed that residents tell their visiting friends and relatives "we have the best/tenth best, etc." garden in the USA.

[3] A hashtag is a word or an unspaced phrase prefixed with a hash character or number sign #. Words or phrases on social media such as Facebook, Instagram, and Twitter may be tagged by entering # before them. A hashtag allows a grouping of similar tagged messages and also allows an electronic search to return all messages that contain it. The term "hashtag" can also refer to the symbol itself and because of its widespread use was added to the *Oxford English Dictionary* in 2014 (https://en.wikipedia.org/wiki/Hashtag, accessed August 4, 2020).

[4] There is also anecdotal evidence that Facebook *groups* are now greatly influencing the choices for garden visiting.

[5] It works on the desktop too, but in a browser. The user can do everything except post but can like, comment, and respond to messages.

[6] Google "McDonald's Millennial Problem" to show the problems a major corporation like McDonald's is having attracting Millennials.

[7] Available at: http://www.blogmetrics.org/gardening (accessed August 10, 2020).

[8] Not coincidentally these three women are perhaps the most respected garden authors on social media today.

[9] It is the nature of blogs that authors come and go and therefore the listing of the popularity of certain blogs changes frequently. Thus in 2015 the most popular blogs in the UK were:

1. Alternative Eden.
2. Fennel & Fern.
3. The Galloping Gardener.
4. Wellywoman.
5. Two Thirsty Gardeners.
6. The Middle-Sized Garden.
7. Real Men Sow.
8. Veg Plotting.
9. The Patient Gardener's Weblog.

[10] Available at: https://www.vuelio.com/uk/social-media-index/top-10-uk-gardening-blogs/ (accessed August 10, 2020).

[11] Available at: http://gardenrant.com/2008/01/who-are-you-p-1.html (accessed August 10, 2020).

[12] E. Licata, 2019, New York, personal communication.

[13] Tumblr, another social medium, also had some strong years between its founding in 2007 and 2018, but in 2019 the average monthly volume of traffic to the Tumblr login page by US visitors dropped by 49% and the average number of daily active users on Tumblr's Android app dropped by 35%, making its survival threatened. Although it claimed to mirror users' personal preferences and hobbies, few gardens adopted the medium. Today, a number of gardens when posting to Instagram will also post to Tumblr given it is only one more click.

[14] See Chapter 7 and the impact of the video by Buffalo Convention and Visitor Bureau on the Buffalo Garden Walk event.

[15] In contrast, in Britain the audience is about 50/50 male to female, while in Japan and North America the female audience predominates.

[16] Tripadvisor is the second most visited travel site in the world (after booking.com, which is part of the Priceline group) with an average of 260 million unique visitors per month. Unfortunately, many recipients of Tripadvisor reports, especially it seems hoteliers, resort to Tripadvisor fraud – posting fake positive reviews themselves or encouraging their personal network to do so – in an effort to boost the hotel's Tripadvisor reputation.

[17] "Solus email" is a term used in the email marketing industry. A solus email is a third-party advertisement that is sent to an opt-in subscription email list that is comprised of demographics of target consumers that the third party wants to advertise to. In the Eden example, *Gardeners Weekly* supplied the list.

[18] Titan arum has the smell of rotting flesh when blooming ... no joke!

[19] See https://www.youtube.com/watch?v=q259TFhPij4 (accessed July 21, 2020).

[20] Pokémon GO is a 2016 augmented-reality mobile game that uses a mobile phone app to direct players to PokéStops (where there are Pokémon characters to be found) and Gyms (which are battle sites) and the locations are typically at places of interest – such as botanic gardens.

[21] Much like Tripadvisor, gardens may use Google My Business to check on their marketplace image. A recent program at Lewis Ginter Botanical Garden received a rating of 4.7 out of 5, thus validating, in the garden's mind, the quality of the program.

[22] Often in competition with the National Geographic Society, the BBC, and other major producers of Webby Award-nominated videos.

[23] The Webby Awards, established in 1997, is the Internet's version of the Oscars and is voted on by over 4 million users. Voters can vote on several categories, of which *Plant One on Me* won in the lifestyle category.

[24] Since 2017 several other garden-dedicated channels have emerged – none with the following of Ms. Rayne Oakes.

[25] Up to 45 min in length but more often less than 30 min; in contrast, her *How To* videos are generally no more than 10 min.

Frida Kahlo art exhibit, Tucson Botanic Garden. Media coverage for posting on social media. Photo reproduced by kind permission of the Tucson Botanic Garden.

"Summer Creatures" come to Newfields, Indianapolis, Illinois in 2018. Posting on social media encouraged. Author's own photo.

The Hawaiiplan, a local Oahu Hawaii travel organiser provides information on best Instagram sites for selfies.

Signs on road in Ho'omaluhia Botanical Gardens. Author's own photo.

Ho'omaluhia Botanical Gardens in Kaneohe, North Shore Oahu Hawaii. Author's own photo.

Skateboarder in Ho'omaluhia Botanical Gardens. Author's own photo.

Montreal Botanical Garden, Quebec, Canada is one of the most important gardens in the world owing to its plant collection. In addition, thematic gardens, such as the First Nations Garden, the Alpine garden, and the Japanese garden, shown above, are unique for the specialized educational programs delivered as part of the garden's mission. Author's own photo.

6

Tourists in the Garden; Human Health and Happiness and the Semiotics of Garden Visiting

DOI: 10.1079/9781789241761.0006

In the previous edition, Chapter 8 was dedicated to tourist motivations when visiting a garden. In this chapter the motivations of garden visiting are explored at a deeper level, into the realm of psychology and psychological drivers to gardens. This research area is called "semiotics".

In the 50 years or so that tourism research has been conducted, research has moved from simplistic descriptions of destinations and participants to the realization that tourist consumption is based on a complex system of motivations, past histories, memories, feelings, desires, and choices that ultimately lead to a level of personal fulfillment and satisfaction. Once these system elements have been acquired in an activity or destination, the participant then has meaning to his or her experience. This is the study of tourism semiotics. To date, little research has been conducted on tourist semiotics although the literature is beginning to explore the dimensions and methodology of this area of tourism research. It is suggested in this chapter that examination of the five senses would be a major indicator of semiotics as much of what a tourist experiences or displays would be recorded through the five human senses. Moreover, the five senses lead to a realization of happiness which is what tourism seeks and indeed what the human species ultimately wishes. Gardens would appear to be an excellent venue in which to examine semiotics as the visitor to the garden uses multiple human senses as conduits to satisfaction and ultimately happiness. Examination of the limited literature on garden tourism suggests that this activity is also a fruitful area for investigation because of the size of the activity in the world today, the nature of the activity insomuch as it involves most of the human senses, and the relative ease of documenting the facets of the tourism activity during its consumption.

Thus, a major regional botanic garden in New York was chosen at which semiotics research might be attempted. The research was primarily that of participant observation, supplemented by a survey probing the motivations for visiting. The results suggest that use of the five senses by visitors – seeing, smelling, taste, hearing, and touching – were major elements in their garden visit. Examination of photography of the garden and the associated plants and flowers suggest a strong reflective attachment to the visit, whereby the visitors searched for the most iconic or classic sight. The major conclusions from this initial research were that participant observation can be a significant tool in understanding tourist characteristics and motivations and that while gardens may have missions that promote activities like education, plant research, and environmental sustainability, tourists in the garden often exhibit behaviors based on individual personal desires that may run contrary to the realization of these goals.

The Nature–Culture Hybrid and Gardens

For at least the past 100 years, nature and culture have been set apart as two distinct ontological tracks with the human set a pole apart from the non-human or nature. This approach was seriously questioned by E.O. Wilson (1984) in his seminal work, *Biophilia*, and given a philosophical grounding in the work of Latour (1991), in which he stresses that the human–nature dualism should be "seen and studied as a series of hybrids made and scrutinized by the public interaction of people, things and concepts". Fig. 6.1 indicates graphically the new postmodernist way of viewing this dichotomy with the work of what Latour calls "purification", the distillation of the nature–cultural divide, and the work of "translation" in the recognition of these hybrids.

Gardens would seem to represent vital and excellent venues in which to study these hybrids because gardens, albeit constructed representations of nature, are areas where humans and their culture come together in a meaningful way. Latour's book, *We Have Never Been Modern*, suggests that indigenous peoples have much to offer this debate as their belief systems and culture have always been a hybrid between their natural world and their cultural traits and beliefs. This link or relationships between gardens and indigenous peoples, unfortunately, is only recent and while it is a fruitful area for new directions in gardens, much research is needed to provide the kind of purification and translation that Latour advocates. The first indigenous peoples' garden in Montreal,

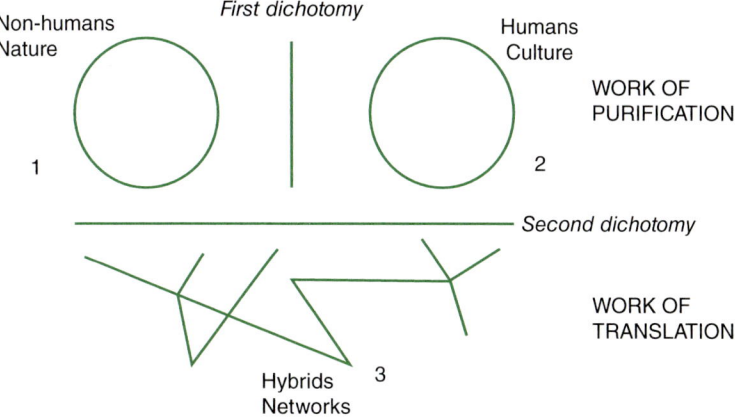

Fig. 6.1. The man–nature interface. After Latour (1991).

Canada opened in 2001, but gardens like the Arizona-Sonora Desert Museum have collaborated on a wide range of projects since 1952. The goal of this collaboration was to "bring together indigenous peoples and western science educators to begin building relationships and exploring the topic of traditional ecological knowledge and western science". In 2019 the APGA began to chart such collaborations to "transform our gardens from being places to showcase our plants ... from our western point of view into places that provide the cultural context for understanding these same plants from the point of view of our land's original inhabitants" (APGA, 2019).

Such collaborations are not confined to North America. In several Australian gardens, Aboriginal gardens and input to programs have been widely developed[1] and in New Zealand, Hamilton Gardens has a dedicated garden built and designed by Māori and used extensively for ceremonies and education. In Mexico, Vallarta Botanical Gardens has included indigenous peoples in many programs and events, and they have even addressed the directors of the APGA Large Gardens. Victor Hugo "Tonatiuh" expressed their goal, which was to: "Share our cultural beliefs focused on maintaining the sacred balance of life/nature ... to preserve peace and natural order ... and through dialogue to understand a collective vision for a world that honors life in all forms and seeks to renew the balance necessary for a prosperous future for all."

History of Psychological Research in Tourism

In the early years of tourism practice and research, the body of work was concerned with tourist numbers, destinations visited, and (economic) outcomes. Indeed, in the tourism industry, the rote collection of destinations that were visited and ticked off a list was summed up in the old industry phrase "if today is Thursday we must be in Brussels". In many ways this phrase encapsulates the earliest research in tourism of linking place and time in a set leisure environment and pattern. However, as early as the mid-1970s (MacCannell, 1976), and some 15 years later (Urry, 1990), researchers suggested that tourists in a certain place were not unaffected *tabula rasa* but came as visitors with complex motivations, choice, movement, feelings, desires, and a personal history that created memories. These human traits are studied under the field of semiotics, which is the study of meaning-making. This includes the study of signs and sign processes (often called "semiosis" or attachment of meaning), indication, designation, likeness, analogy, metaphor, symbolism, signification, and communication. It is believed that if we understand these human traits, a more detailed appreciation of tourist attractions and the associated visitor management might be garnered, thus leading to understanding and explanation in tourist activity needs, wants, and desires.

Since Urry's seminal work, semiotic research has been focused on the *visual* manifestations of the detailed psychology of the tourist or visitor, for tourism is inherently about the visual (photography, brochures, maps, postcards), and over the succeeding past 30 years there has been a significant body of theoretical work informing tourist semiotics but little applied research in this field. Alongside this, more recently, there have been cries from tourism researchers for a wider perspective on semiotics that considers not just the seen or visual but "… the desire and anticipation that influences behavior (Pritchard and Morgan 2010), motivation and choice to movement … wandering gazing, exploring … feeling light and shade, happiness, gloom … indeed there is a whole repertoire of … *sense registers* [author's italics] rooted in framing the tourist experience" (Waterton and Watson, 2014). This chapter is an early attempt at exploration of tourism semiotics in the field of garden tourism; the earlier research by the author (Benfield, 2013) suggested that garden tourism is one of the most important areas for the study of semiotics because it does not restrict itself to just the visual but also involves the sense registers noted above concealing deeper emotions, for garden tourists' activities and patterns invariably are involved in seeing (the visual), as well as sounds, tastes, smells, and touch.

While there is a paucity of literature on the specific application of semiotics to garden tourism, there is a wide body of literature from the psychological and sociological sciences upon which one may infer relationships to tourist travel motivations. The earliest literature on the sociology of travel and tourism originates with Boorstin (1964), who places the tourist within an environment of inauthentic, contrived group travel, enjoying what he called "pseudo events". Turner and Ash (1975) were the first to recognize this as a mass phenomenon and derived the term "mass tourism" to describe staged authenticity for a mass of gullible and vacuous individuals being shown a restricted, scripted, and thus sanitized version of what was reality. Thus, as noted above, it was MacCannell, with the publication of *The Tourist* in 1976, who moved away from the explanation of a mass of gullible individuals but rather suggested that individuals were motivated by what resembled a sacred journey or pilgrimage, looking for authenticity (which was a reaction to what he identified as "staged authenticity") in their tourist travels and rejecting staged authenticity. John Urry's work in 1990 indicated that the major feature of this viewing of "the other" was the visual and hence termed the phenomenon "the tourist gaze". Both must be seen, therefore, as the earliest and most impactful publications on tourist semiotics. Since 1990, Urry's *The Tourist Gaze* has been reissued as a third edition (Urry and Larsen, 2011); some of the findings are new (especially in the area of globalization and performance – see below) and some of the old bear repetition. First and foremost, Urry and Larsen (2011) tie the tourist gaze into the search for pleasure and happiness when people leave their normal abode and travel. Furthermore, the authors note "the notion of the tourist gaze is not meant to account for why specific individuals are motivated to travel" (Urry and Larsen, 2011); rather gazes are related to building, design, content, and social practices and discourses. Finally, rather than the static gaze of viewing or visualizing, Urry and Larsen (2011) now suggest there is a less static gaze called the "performance turn" where "tourism demands new metaphors based more on being, *doing, touching, and seeing* [author's italics] rather than just 'seeing'" (Urry and Larsen, 2011). Thus, in the progression of this social science research to garden tourism, it is suggested, sequentially, that:

- Individuals come to gardens – tourism.
- According to Urry, tourist gazes are structured according to class, gender, age, and ethnicity. Thus, in any investigation of semiotics, demographic variables become important to frame the discussion that ensues.
- Tourists travel to find pleasure and happiness.
- Tourists come to gardens for pleasure and happiness (possibly as opposed to casinos, theme parks, etc.).
- Tourists come with differing motivations.
- What they participate in, in the garden, is related to garden building, design, and content.

- The researcher might expect to encounter social practices and discourses in the garden.
- Gazing is a set of practices that can be seen, probed, and displayed.

What is Happiness? And its Relationship to Garden Tourism

If we accept Urry's thesis that when people leave their normal abode and travel, the tourist gazes to search for pleasure and happiness, then it is axiomatic that we should begin by investigating the nature of happiness.[2] The 19th century philosopher, Jeremy Bentham, one of the most famous enlightenment thinkers and moral philosophers, espoused his doctrine of utilitarianism which suggested that the relevant consequence of action is the creation of overall happiness for everyone who is affected by the action (Sweet, n.d.). Bentham developed this theory by grounding it in a largely empiricist account of human nature. He famously held a hedonistic account of both motivation and value according to which, what is fundamentally valuable and what ultimately motivates us, is pleasure and pain (Sweet, n.d.). Happiness, according to Bentham, is thus a matter of seeking and experiencing pleasure, and it is a valid extension to suggest that garden visiting is thus a search for happiness. Wikipedia (n.d.) defines happiness as a mental or emotional state of well-being characterized by positive or pleasant emotions ranging from contentment to intense joy. A variety of biological, psychological, religious, and philosophical approaches have striven to define happiness and identify its sources. Seligman (2002) asserts that happiness is not solely derived from external, momentary pleasures ("the gaze"?), and humans seem happiest when they have:

1. Pleasure (tasty food, warm baths, etc.).
2. Engagement (or flow – the absorption of an enjoyed yet challenging activity).
3. Relationships (social ties have turned out to be an extremely reliable indicator of happiness).
4. Meaning (a perceived quest or belonging to something bigger).
5. Accomplishments (having realized tangible goals).

In this regard gardens represent a major opportunity for provision of the first four of Seligman's five items, and particularly the first – pleasure. Wilson (1984) was one of the first to link the happiness of humans to biological species; and regarding pleasure, the five human senses would seem the most obvious receptors to derive that pleasure in the context of plant species. The field of perceptual psychology based on the stimulation of senses has much to offer in that regard and the works of Rosenblum (2010), Wysocki and Preti (2002), and Jacob and McClintock (2000) are particularly noteworthy in understanding the link between senses and human social communication.[3] Morton (2014) indicates all five senses are important to flower gazing and garden visiting:

> Flowers are good for psychological and physical health. A flower garden offers experience of all five senses: touch, taste, smell, hearing and seeing. This culminates in a state of enjoyment. A walk through a garden of flowers expands and enriches nerve cells in the brain that help create feelings of contentment and cooperation – enjoyment and it has been shown in numerous studies, feelings of enjoyment enhance health. Each sense has a linkage to garden enjoyment:
>
> **Touch**
> Touching a flower activates the body's sense receptors that provide information to the brain. Touch determinates safety or danger and stimulates a hormone response appropriate to either situation. Flowers are often soft to the touch, which enhances the feeling of safety and pleasure. Gentle touch releases the hormone oxytocin, which is responsible for feelings of trust and cooperation.
>
> **Taste**
> Marigolds, nasturtiums and roses are flowers that are often used in salads. Nasturtiums have a sharpness that stimulates the taste buds on the back of the tongue. The taste buds that detect the sweetness of rose petals are located on the front of the tongue. Anise plants produce a strong licorice taste that the sense of taste perceives as soothing and sweet.
>
> **Smell**
> According to Rosenblum (2010), smelling activates nerve cells that induce brain activity related to memory, attention and emotion. There is no quicker way to enhance feelings of contentment than by smelling flowers that you love. Memories of earlier pleasures are

stimulated, releasing hormones that produce contentment. The aroma of essential oil molecules works through hormone like chemicals to produce results. Scents and odors influence the glands responsible for hormone levels, metabolism, insulin, stress levels, appetite, body temperature, and even sex drive. Actual studies of brain waves show that scents like lavender increase alpha brain waves (associated with relaxation) (see Fig. 6.2).

Sound

Flowers are often accompanied by the sound of birds singing and bees buzzing. Nerve cells in the inner ear send signals to the brain that let us know whether a sound is disturbing or pleasurable. The brain reacts by creating good feelings when the sound is a familiar good one, like birds singing near flowers. Pleasant sounds relax the nervous system and helps maintain physical and emotional balance.

Sight

The sight of beautiful flowers stimulates emotional reactions of contentment. Eyes perceive texture, form, movement, light and shadow in a garden, each experience sending information to the brain. Pleasure centers are activated by beautiful flowers and feelings of enjoyment are the result. The experience of pleasure from flowers is enhanced when accompanied by other sensory inputs such as the sound of a bubbling fountain.

(Morton, 2014)

The Importance of Gardens to Human Health

Prior to looking at the patterns of movement and actions of tourists in a garden, it is appropriate to reference the literature that links gardens (and gardening) to human health and well-being.

There are three main theories linking human health with the natural environment:

- biophilia theory – that humans seek contact with other species and have a need to be close to nature (Croucher *et al.*, 2008);
- attention restorative theory – that the outdoors and natural environments can assist in recovery from attention fatigue by allowing people to distance themselves from routine activities (Croucher *et al.*, 2008; Nordh *et al.*, 2009); and
- stress reduction theory – that the outdoors can assist in recovery from stressful events[4] by different types of environments triggering emotional and physiological responses (Ulrich, 1983; Croucher *et al.*, 2008; Nordh *et al.*, 2009).

What is most important is that all three of these theories propose some deep genetic coding that underpins our preference and ability to recover in natural environments (Depledge *et al.*, 2011).

Since the work of the early years of the 21st century, several studies have concentrated on individual plants and environments to chart the effect of gardens on health. In 2014, Hartig *et al.* isolated nature as benefiting health in the areas of less stress, more exercise, more positive social interactions, and better quality of environment for health promotion. Lin *et al.* (2014) showed how trees strongly and positively affect cognitive performance and Beyer *et al.* (2014) showed how the provision of trees and vegetation in general reduced both depression and stress. Furthermore,

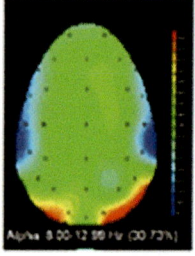
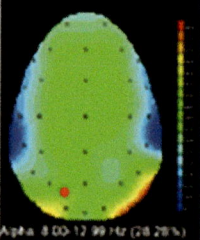
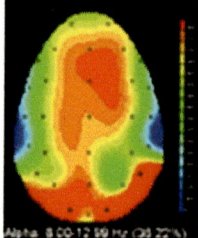

Eyes closed   Sweet almond oil inhalation   Lavender inhalation

Fig. 6.2. Effects on the human brain of inhalation of almond and lavender smells. Brain topographical map of the distribution of alpha brain-wave activity, where red indicates a significant increase in power in bilateral temporal and central areas during inhalation of lavender. After Sayorwan *et al.* (2012).

work by Mitchell and Popham (2008) showed that green space reduced social inequalities between income groups.

Much of it was summarized and brought into the full context of health and gardens by a symposium entitled "Health and Horticulture" held at the garden of Hampton Court Palace in London in July 2016, at which Cameron and Griffiths provided the clearest work on the link between gardens and gardening and individual health. Table 6.1 (Cameron and Griffiths, 2016) indicates the reported and perceived benefits of green space, gardens, and gardening.[5]

Within the garden, Ivarsson and Hagerhall (2008) and Ogunseitan (2005) provide the most detailed insights on gardens' effects. Ivarsson and Hagerhall found from three gardens in Sweden that natural features provide the greatest relief from stress while Ogunseitan found that within the natural features,[6] preferences for flowers and water were associated with high quality of life.[7] In addition, Ogunseitan's research showed the importance colors and sounds have in the synesthetic[8] tendency of the environment.

Finally, Cameron and Griffiths (2016) point out that while gardens and gardening are under-researched in the nature–health agenda, the benefits could potentially be very large.[9] Furthermore, one of the most important research areas is the examination of landscapes. It appears that heterogeneous landscapes are the most important type of landscape not only for typology but also uses and attitudes. In this regard the design of gardens, particularly botanic gardens, warrants further investigation as it may be the most important factor in stress relief from work, commuting, and social/family life.

The link between health and garden tourism

In 2006 the Government of Ontario studied American (and Canadian) overnight travelers to ascertain the benefits and motivations for undertaking a vacation. Perhaps remarkably, at the top of this list was the statement that a vacation was used primarily to relieve stress – 85% of all American travelers used travel to relieve stress in one form or another – while the second most important reason, stated by 69%, was to maintain or improve personal relationships. Only 54% of travelers reported that they went on vacation to seek knowledge and mental stimulation.

Activity participation at home

American adults (travelers and non-travelers) say that their preferred outdoor activities at home are outings to parks, swimming, exercising, and gardening, with gardening second only to day trips to parks as preferred activities. When asked to be more specific, almost 20% of Americans visit a botanic garden at least once per year or even more frequently.

Garden research and applied semiotics

Many have noted that gardening and garden viewing is all about the "being, doing, touching, and seeing" as described by Urry and Larsen (2011). Thus, garden tourism would represent a significant activity that might be studied for the role of semiotics in one, and perhaps other, tourism activities.

In 2004, Joanne Connell wrote the first and seminal work on garden visitors (Connell, 2004) and while her initial objective was to provide

Table 6.1. Reported and derived benefits of green space, gardens, and gardening. After Cameron and Griffiths (2016).

Green space	Gardens	Gardening offsets …
Pain relief	Reduced mortality	Heart disease
Blood pressure	Higher bone density and less osteoporosis	Ischemic stroke
Heart rate	Cholesterol levels	Type 2 diabetes
Less frequent illness	Reduces onset of dementia	Hypertension
Improved cognitive function		Anxiety and depression
Thermal comfort		Certain types of cancer

baseline data on the previously completely ignored activity of garden visiting, she focused not
only on the characteristics of garden visitors but
also on their motivations. To explain motivation
Connell undertook a factor analysis and remarkably she found that three components explained
65% of the variation in the data. The three dimensions of garden visitor motivation are:

- social (28%) – enjoying the company of
  others (family, group);
- horticultural (21%) – related to their own
  garden; and
- setting (16%) – sensual emersion, peace,
  tranquility.

Clearly social and setting, at almost two-thirds
of the components identified, suggest that semiotics is a vital and fruitful area of research in
understanding garden tourist visitation. Connell
also provided cross-tabulation of the visitors'
age, occupational class, and typology with reference to their activities while in the garden. The
results are shown as Table 6.2.

What it clearly indicated was those more
sedentary activities, or what Urry might identify
as social discourse (sitting and chatting), are
undertaken by all age groups, while active activities such as picnicking and painting are
weighted toward the younger age cohorts. Older

visitors are more likely to take notes (for educational purposes?) and undertake photography.
Occupational groups appeared to show little difference in their activities in the garden except
chatting, which was more important in more
manual labor groups, a group that had less
interest also in nature study.

In 2014 Cengage Learning, a college educational company, applied Connell's findings to
Urry's classification of tourist gazes as a case
study. The results are reproduced in Table 6.3. It
provides the marriage between garden visiting
and the semiotics of tourist experiences and can
thus be tested by application of tourist patterns
and activities in a garden.

Following is a case study in which university
students probed, by global positioning system
(GPS) tracking, selected interviews, and survey administration, what visitors did while in a garden.

Case Study 6.1: Queens Botanical Garden, Flushing, New York, USA

The Queens Botanical Garden began as part of
the 1939 New York World's Fair in Queens.
After the fair, the garden expanded to take up a
larger portion of Flushing Meadows Park.
When work was begun on construction of the

Table 6.2. UK garden visit activities (percentage of visitors participating by category). After Connell (2004).

	Photography	Nature study	Painting	Picnicking	Sitting	Chatting	Taking notes
Age (years)							
18–39	43.4	27.7	4.8	44.6	73.5	59.0	25.3
40–60	51.3	43.7	8.0	28.7	76.6	49.0	49.4
Over 60	47.7	34.4	3.6	27.2	72.8	47.2	56.4
P value	0.417	0.015	0.123	0.10	0.624	0.180	0.000
Occupational grading							
AB	44.7	39.6	5.8	28.7	74.5	48.0	48.7
CIC2	55.3	40.2	6.7	30.7	76.5	49.7	47.5
DE	43.9	23.8	9.5	38.1	76.2	71.4	42.9
P value	0.076	0.335	0.769	0.634	0.887	0.117	0.861
Type of visitor							
Special horticultural interest	48.1	51.9	7.4	14.8	61.1	37.0	75.9
General gardening interest	51.2	35.9	5.5	27.1	74.8	47.9	51.2
Just seeking pleasant day out	35.9	35.0	6.8	48.5	83.5	64.1	17.5
P value	0.023	0.066	0.786	0.000	0.008	0.002	0.000

Table 6.3. The tourist gaze and reasons for garden visiting. After Connell (2002).

Reason for visit	Form of tourist gaze	Type of visit
For a day out	Spectatorial	Casual
To enjoy a garden	Romantic	Casual
For interest/curiosity	Spectatorial/romantic	Casual
To see something specific	Spectatorial	Purposive
Been before	Anthropological	Casual
Group visit	Collective/spectatorial	Purposive
To show someone else	Collective	Purposive
To get ideas	Spectatorial/romantic	Purposive
To see progress	Anthropological/environmental	Purposive
Magazine/television feature	Spectatorial	Purposive
Yellow Book	Spectatorial	Purposive
Saw leaflet	Spectatorial	Purposive

1964 World's Fair, the garden was moved across the street from Flushing Meadows Park to a site located atop the stream bed of Kissena Creek, a water source that is now a focal point within the garden. Since 1964 the residential area around the garden has undergone significant change, moving from a higher-income, predominantly white neighborhood to a more multi-residential area and multi-ethnic community, making it one of the most ethnically diverse counties in the USA. Approximately 150 nations are represented in the borough. The rich cultural traditions of the immigrant and ethnic communities that have made Queens their home reverberate on its streets. In Flushing alone, Indian, Afghan, Chinese, and Korean cultures, to name a few, mingle with one another within a small area of a few blocks and have become a permanent part of the Queens landscape. All these ethnic groups use the garden for leisure activities, often because green space and recreational areas are in short supply. The garden thus represented a good case study owing to the cultural and demographic diversity coming to the garden. The garden is also a tourist attraction for visitors to the greater New York area.

The Queens Botanical Garden itself now consists of 39 acres (158,000 m²) of rose, bee, herb, and perennial gardens. It is open to the public and a fee for entry of US$4.00 was initiated in 2012 to offset projected deficits. Residents can enjoy the garden by purchasing a membership and gain unlimited access. Wedding photography is popular in a special wedding garden and on the lawns by appointment.

Methodology of the semiotic research

On October 10, 2009, visitors to the garden were researched – primarily using participant observation – to document the signs they displayed during their leisure time in the garden. The following techniques were used:

- GPS. A random selection of visitors was discreetly followed as they walked the garden, charting where they stopped and the activities they pursued during the walk (sitting, talking, smelling flowers, touching artifacts).
- Video. Individuals were videotaped, with their permission, as they walked within the garden and a semi-structured interview was administered.
- Semi-structured interview. Individuals were requested to respond on video to the following questions:
  - their history of visiting Queens Botanical Garden;
  - why they were there today; and
  - why they choose a garden and particularly Queens Botanical Garden.
- Photographs. Photographs were taken of every individual activity undertaken in the garden, both active and sedentary.
- Survey. Garden visitors were surveyed on entry to the garden using a Likert scale

from 1 ("low") to 5 ("high") to ascertain their motivations for visiting the garden that day. The motivational areas probed for coming to the garden were:

- to see just flowers;
- to learn about plants;
- to see a garden;
- to enjoy the garden environment;
- to learn about gardening (educational motivation);
- to generally learn about something;
- to get exercise; and
- to see the building.

Case study findings by methodology used

GPS

Fig. 6.3 shows the pattern of movement traced by a selection of visitors using a GPS unit. Blue flags indicate where they stopped for a period (also timed) while the red flags indicate locations of signage. The figure shows a set circular pattern to movement around the periphery of the garden and along internal arterials. There are certain congregational points that are highly significant. These are:

- benches along major arterials;
- locations of significant floral splendor;

- locations where the water is moving, thus creating sparkling, reflective images and babbling, pleasurable noise; and
- the wedding area (people congregating even when there was no ceremony in progress).

Areas of significant avoidance appear to be:

- signage locations;
- the more open area of woody species, although some people will leave the path to examine individual trees and shrubs; and
- the center of the grassed area (people tend to sit on the grass in proximity to plant species rather than away from the plants).

ACTIVITIES. Recorded leisure activities in the garden were:

- Strolling and walking (most visitors):
  - sitting on benches and talking;
  - smelling flowers;
  - touching plant species; and
  - looking at flowers.
- Reading.
- Tai Chi.
- Lying on grass.
- Sitting on grass while children play.
- Photography.
- Wedding:
  - wedding photography; and
  - standing around chatting/reconnecting with friends/family.

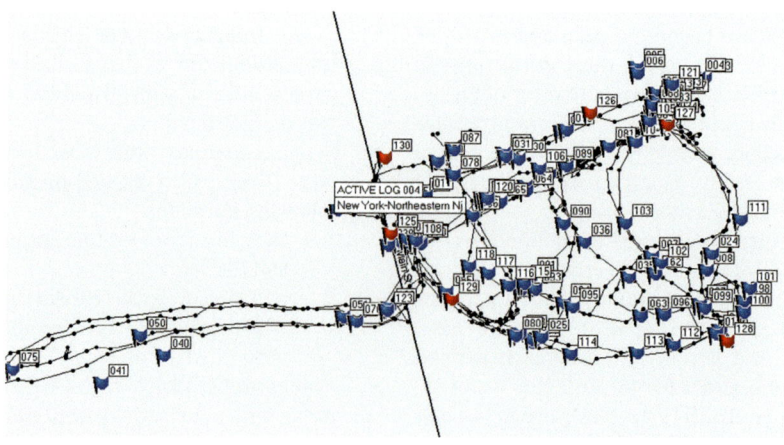

Fig. 6.3. Example of GPS monitoring of a group of two visitors at Queens Botanic Garden, Flushing, New York, in fall 2009. Blue flags are stops; red flags are location of signage.

Semi-structured interviews

In response to the three areas probed by the questions, visitors to the garden were invariably frequent visitors to Queens Botanical Garden, many coming on a daily or weekly basis. They had been visitors for many years and, as is seen below, had a strong personal attachment to or "ownership" of the garden. Some had been brought up in the neighborhood and were now getting married in the garden. The garden was part of their perceptual geography:

- many came for the events in the garden (Diwali Festival, marriages);
- many came as a place to walk away from traffic;
- many came to pursue sedentary activities (reading, sleeping, picnicking, and sitting and doing nothing); and
- many came to pursue hobbies (photography).

When probed as to why they chose this garden – Queens Botanical Garden – most said because of the welcoming nature of the administration.

Survey

Table 6.4 presents detailed survey results.

Photography

Because there was a limited number of people in the garden who were seen with cameras on the two days the research was undertaken and there was no way the photographs taken by the research subjects could be obtained, a surrogate sample of 162 photographs taken in Brooklyn Botanic Garden[10] and downloaded from Flickr

Table 6.4. Reasons for garden visiting. Queens Botanical Garden, Flushing, New York, 2009.

Motivator	Mean score	Comments
To enjoy the garden environment	4.66	It is clear the ambience and environment offered by a garden was the principal reason for tourist visitation
To see just flowers	4.10	The presence of flowers as THE visual attraction is suggested, but possibly the other sensual attractions are strong motivators for visitation
To generally learn about something	3.52	Connell suggested that horticultural education is a significant motivator to garden tourism, but this finding reveals that education in a wider context (environmental education, botanical discovery, etc.) may also be a prime motivator
To get exercise	3.35	When walking is the most popular outdoor activity of Americans today, the attraction of a free and pleasurable open space for walking (and Tai Chi) seems evident by this finding
To learn about plants	2.98	As Connell suggested that horticultural education is a significant motivator to garden visitation, this finding would suggest that Queens Botanical Garden is fulfilling this role in a significant way
To see the building	2.67	The administration building in the garden had recently been designated a LEED building[a]. Some may have come just to see what such a designation entailed. Others may have responded positively to this question as events were being held in the building that Saturday morning
To learn about gardening	2.39	The fact that this was a fine weather day and no gardening education courses were being offered that day suggests that this motivator was low on motivational reasons
To see a garden	2.05	Queens Botanical Garden contains rose, bee, herb, woody species, wedding, and perennial gardens. It appeared few came specifically for a particular garden

[a] LEED (Leadership in Energy and Environmental Design) is the most widely used green building rating system in the world. Available for virtually all building types, LEED provides a framework for healthy, highly efficient, and cost-saving green buildings (https://www.usgbc.org/help/what-leed, accessed August 13, 2020).

were analyzed for content. The results were as follows:

Photographs of individual flowers	58.02%
Photographs of floral landscapes	16.07%
Photographs of floral landscapes with people in the photograph	9.25%
Photographs of animals, insects, and birds	5.55%
Photographs of individual persons	4.32%
Other photographs (signs, entertainers)	6.79%

Waterton and Watson (2014) suggest that tourism photographs "are felt and experienced, they are not really taken ... what we see here is that what we are looking at is a semiotic that extends from the visual into other senses literally *making sense of place*" (Waterton and Watson's italics). Accordingly, they divide photographs into three orders:

- the reflective;
- the imaginative; and
- the immersive.

In the case of the botanic garden, as is the case in heritage sites, the reflective – characterized by the "standard, classic, even iconic" – is the most photographed (flowers were featured 58% of the time), the affective (creation of moods and memories, imaginative) is less well represented with 16%, while the immersive or personal – in which the individual becomes one with the product – is very rarely found (4%).

Conclusions, Implications, and Future Research

Visitors to botanic gardens are primarily motivated to visit by two features: social reasons and the setting. These motivations are not mutually exclusive and in fact evidence from viewing activities in the garden would suggest they are heavily related. At the social level, the visitor who comes for non-educational reasons is highly desirous of an environment that permits social discourse with whomever they are with. They use the infrastructure heavily to meet this social need by sitting on benches and chatting or sitting on the grass and picnicking or watching their children play. Concerning the setting, the fact that 83% come to a garden for the flowers is borne out by the reflective photography that is taken and suggests that the garden is delivering significant but varied meaning to all visitors. When investigating the use of sensory stimuli, the fact that the visitors stop at the places of significant floral splendor or noise from running water, and they smell, touch, and listen, means that their gaze is directed at what they come to see. This has serious implications for garden management because most botanic gardens have as their mission:

- education;[11]
- sustainability; and
- research.

However, as shown by their sensory habits and actions, particularly seeing but also hearing and touching, the visitors surveyed at Queens Botanical Garden come to the garden for:

- stress release;
- family time; and
- relaxation.

Thus, for example, signage in the garden that promotes environmental issues and causes, botanical science, or is just generally educational in nature is ignored and is thus ineffective.

Clearly this body of work needs refinement. Additional research might involve:

- Cross-tabulating activities ("the gaze") with age, income, employment, and ethnicity.
- Acquiring detailed personal histories of the participants who are coming to the garden to obtain a record of memories and reflections.
- Exploring the detailed reactions of the participants when they smell a flower, see a bird, hear a bird, touch a petal, or taste a plant.
- It is of singular interest to the author as to why garden visitors respond so dramatically to the presence of water. Many persons stopped at the water with no prompting or ostensible reason to stop; numerous groups gathered around the bridge over Kissena Creek, children would stop to touch the water flow, and it was photographed numerous times by visitors. What are the properties of water that make the visitor stop and presumably happy?
- Why in photography does the tourist stress the reflective and not the immersive? Surely

the latter is more meaningful to the individual in terms of self-actualization?

Finally, while we might conclude and confirm Francis Bacon's contention that garden visiting is indeed the purest of human pleasures, there is much yet to understand both emotionally and psychologically as to *why* gardens and flowers generate so much pleasure and happiness.

Notes

[1] See *Garden Tourism* (Benfield, 2013) and the Cadi Jam Ora garden at the Royal Botanic Garden, Sydney.

[2] The study of human happiness is certainly an emerging new direction in the world. In 2011 the United Nations began to rank national happiness in the world and today a National Happiness Index is published yearly for all countries. In the USA, the promotion of happiness is being led by https://www.happycounts.org/ (accessed July 22, 2020). A small number of researchers in the USA and Europe are applying this to food and plant interactions.

[3] Nadel-Klein (2010b), in her work "Gardening in time: happiness and memory in American horticulture", provides the definitive anthropological link to gardening and happiness with direct references to memory, emotion, senses, and their connections – all key study areas in the semiotics of gardens and thus garden tourism.

[4] In the only study of its kind, Kohlleppel *et al*. (2001) provide a direct link between a walk in a botanic garden and the reduction of stress.

[5] Cameron and Griffiths (2016) note that the extent of these benefits may relate to the individual's attitude to gardening (and gardens?).

[6] Ogunseitan used the concept of topophilia (Tuan, 1974) to characterize the restorative environment of a garden. Topohilia – the affective bond between people and place of environmental setting – is broken down into four categories: synthetic tendency, environmental familiarity, cognitive challenge, and eco diversity.

[7] It was under eco diversity that Ogunseitan found the remarkable effect of water and trees.

[8] Synesthetic tendency is the comingling of sensory stimulations and the memory of place, thus reinforcing the importance of the sense stimulation of gardens.

[9] It is worth noting that Cameron and Griffiths (2016) isolate the attention restorative theory (see above) as the most important function of a garden for stress relief.

[10] Located about 5 miles distant from Queens Botanical Garden.

[11] Williams *et al*. (2015) examined in detail the question of whether botanic gardens can positively influence visitors' ecological knowledge and environmental attitudes. They showed that botanic gardens have very little influence on knowledge, but positively influence attitudes toward the environment.

Kings Park and Botanic Garden, Perth, Australia promotes the garden as a refuge from stress. Author's own photo.

Signs in Kew Gardens encourage guests to smell the flora. Author's own photo.

In Helen's Garden for the Senses in New York Botanical Garden, guests are encouraged to touch the plants for their tactile properties. Author's own photo.

Patrons touch the plants for their sensory properties in Helen's Garden for the Senses, New York Botanical Garden. Author's own photo.

Helen's Garden for the Senses in New York Botanical Garden is designed and planted specifically for sensory experiences. Author's own photo.

Braille trail, Lowveld Botanical Garden, Nelspruit, South Africa. Photo courtesy of Lou-Nita Le Roux, SANBI.

The University of Connecticut Cognitive Garden at Avery Point attempts to engage users, particularly children, in sensory stimulation and experiential learning. Photo courtesy of Annette Montoya.

Demonstration of how to make a historically accurate tussie-mussiet during a garden conservancy open day at the Gardens at Clock Barn in Carlisle, Massachusetts. Photo by Brian Jones.

The Orchid Show, New York Botanical Garden, The Bronx, New York is the largest event in the garden's event calendar running for six weeks in the spring; it attracts over 250,000 visitors annually. Author's own photo.

7

Events and Festivals

―――――――――――――

© Richard W. Benfield 2021. *New Directions in Garden Tourism* (R. Benfield)
DOI: 10.1079/9781789241761.0007

In discussing the "Progress and prospects for event tourism research", Getz and Page (2016b) state that "growth of this sector … can only be described as spectacular". They go on to chart that, in the period 2006–2014, there may have been over 15,000 articles dedicated to event tourism. Unfortunately, much like in other areas of garden tourism, research on events and gardens is limited. SCOPUS results based on a search by Fox (2017a) revealed over 1190 articles on gardens and tourism, but when the search was refined to include the terms *botanic* and *event* or *tourism*, only two articles appeared. Again, this is an area that is ripe for research and understanding.

Notwithstanding the lack of research on garden events per se, festivals and events are a hugely important direct and indirect drivers of tourism to gardens. As an example, over many years Kew Gardens has built up a visitor program of events and festivals including (Barley, 2019):

- an intergenerational family offer during school half-term and summer holidays with programming that complements Kew's summer festival;
- an annual orchid festival thematically focused round a different nation each year;
- Easter festivals working with mainstream family brands;
- after-hours events running throughout the year, following the seasonal themes;
- Kew Science Festival (every year at Kew and/or Wakehurst);
- programming in the Shirley Sherwood Gallery, highlighting contemporary botanically inspired artists as well as botanical artists;
- the 2019 glass exhibition "Chihuly: Reflections on Nature";
- "Thriving and Surviving" exhibition at Wakehurst;
- Kew The Music;
- Christmas at Kew; and
- new commercial event concepts that take place outside of normal opening hours.

All these offers and special exhibitions drive new audiences and attract repeat visitors. Market research at Kew shows that:

- for 8% of visitors, festivals and events are the main motivation for visiting Kew Gardens, in winter this more than doubles to 19%;

- 13% of total visitors list events and temporary exhibitions as the most enjoyable aspects of their visit to Kew; and
- 74% of visitors to the Annual Orchid Festival said the event had affected their decision to visit Kew Gardens.

Furthermore, research shows that events, festivals, and exhibitions appeal to some target segments more than others do. As a result, Kew uses programming as a part of the marketing mix to target specific audiences; for example, families looking for activities in the school holiday. Programming is also a means of engaging and informing audiences by interpreting collections and enabling Kew to communicate messages about its mission and values, changing perceptions of Kew and creating informal learning opportunities all the while contributing to the domestic and international tourist offer in the UK.

Notwithstanding the paucity of specific published studies, the researcher can use the article by Getz and Page (2016b) as a framework in which to suggest the importance of events in gardens. The fact that all public gardens in the USA, Canada, the EU, and Australasia undertake events and festivals, even though rarely researched, suggests that event tourism in gardens will be a significant part of the future of gardens well into the 21st century.[1]

Getz and Page (2016b) provide a typology of events in the context of event tourism as shown in Fig. 7.1. In the model, while not explicitly noted, gardens are a major venue for these tourism events, but garden tourism also takes place in street festivals, urban parks, and concert areas within a garden area. Indeed, in the typology suggested by Getz and Page, it appears that only sporting events are not found in a garden or green-space setting.

Meetings and conventions, the business category of Getz and Page's (2016b) typology, are often held in gardens, mostly as botanical meetings by plant societies, artists, and garden clubs, but possibly the most important activity is weddings and receptions. Thus, for example, a garden like Queens Botanical Garden in Flushing, New York obtains 15% of its earned income from weddings, catering to up to five weddings per day some days in June in dedicated areas of the garden and afterward in the garden's meeting space (see section on "The importance of weddings" below).

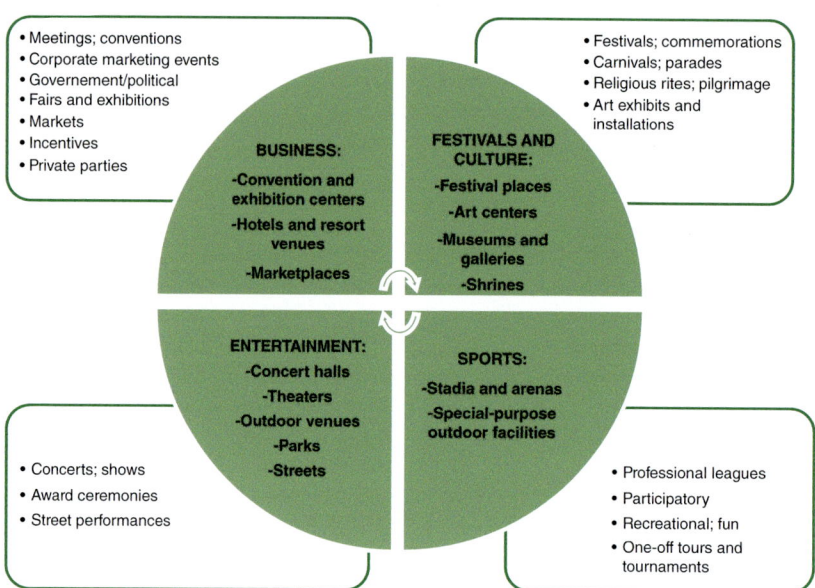

Fig. 7.1. Typology of tourism and hospitality events. Reproduced from Getz and Page (2016a), by arrangement with Taylor & Francis Books UK. © 2016 Donald Getz and Stephen J. Page.

The second category, festivals and culture, is a major activity for garden tourism. Art exhibits and installations, parades, specific topical displays, ethnic festivals, and religious festivals are all part of the event mix in botanic gardens. For example, Queens Botanical Garden holds events for, among others, Indian (Diwali), Burmese, Chinese, and Hispanic guests, particularly on special festival days (see Case Study 7.1).

Entertainment, the third category, has become a major part of the event mix for gardens with symphonic concerts, rock concerts, and less impactful concerts of folk artists, R&B bands, solo artists, and stage plays. The State Botanical Garden of Georgia in Athens provides concerts in the garden every Friday night, attracting 300–400 guests.

Getz and Page (2016b) then differentiate events by size of event, the so-called "portfolio approach" to classifying events (Fig. 7.2).

While much of what Getz and Page (2016b) discuss regarding the progress in event tourism is not yet applicable to garden tourism events owing to the lack of research findings, two significant areas of research that do impact garden events are the discussion on event tourism experiences and the temporal issues in event tourism. In the former, event tourism is directly related to

the experience economy of Pine and Gilmore (1999) (see Chapter 3) and the range of event experiences is broad, ranging all the way from fun and revelry (craft beer events at Norfolk Botanical Garden and Newfields, Indianapolis) to the formalization of rituals and pilgrimage (Diwali at Queens Botanical Garden). Within these experiences one can conceptualize the study of event participation in terms of what people do or how they behave (the "conative" dimension) or what moods, emotions, and attitudes are prevalent during the event (the "affective" dimension); a third conceptualization is the participants' experience in the realm of cognition or awareness, perceptions, and understanding. The second area of research that seems important to gardens is the discussion of seasonality. The research of Connell *et al.* (2015) suggests that gardens might be both typical and atypical in this area. Connell and colleagues point out that most events are scheduled in response to off-peak or reduced demand for a product. Yet most events in gardens take place in the summer high season when demand is greatest. Significantly, in the case of gardens, summer or high season is the time of maximum bloom and thus many gardens concurrently hold their major events during July and August (e.g. Idaho Botanical Garden

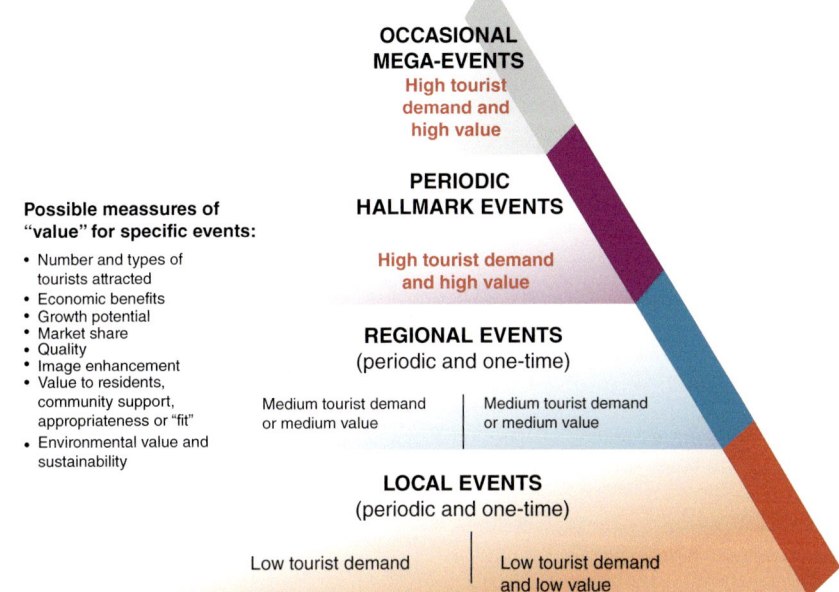

OCCASIONAL MEGA-EVENTS
High tourist demand and high value

PERIODIC HALLMARK EVENTS
High tourist demand and high value

Possible meassures of "value" for specific events:

- Number and types of tourists attracted
- Economic benefits
- Growth potential
- Market share
- Quality
- Image enhancement
- Value to residents, community support, appropriateness or "fit"
- Environmental value and sustainability

REGIONAL EVENTS
(periodic and one-time)

Medium tourist demand or medium value

Medium tourist demand or medium value

LOCAL EVENTS
(periodic and one-time)

Low tourist demand

Low tourist demand and low value

Fig. 7.2. Hierarchy of events: a portfolio of events by type, season, target markets, and value. Reproduced from Getz and Page (2016a), by arrangement with Taylor & Francis Books UK. © 2016 Donald Getz and Stephen J. Page.

in Boise). However, in the growing season or dormant period, events are also deliberately staged to stimulate attendance. It is instructive to look at a typical year of events by one garden to see how it obtains a balanced portfolio of events.

Top events in 2019 by month, Tower Hill Botanic Garden, Worcester, Massachusetts:

- January: closed.
- February: Healers and Killers exhibition, Apothecary in Bloom floral design show, Camellia Show, Apothecary Marketplace, Seeds of Sorcery.
- March: Horticultural Heroes exhibit opens, Let's Get Growing seed starting programs.
- April: Hello Spring exhibition begins, George Sherwood sculpture opening, African Violet Show.
- May: Daffodil Days, Primrose Show, Daffodil Show, Rhododendron Show.
- June: Tinker Garden, Rose Show.
- July: Botanical Tattoo Weekend, Daylily Show.
- August: Caterpillar Show, Kinetic Races.
- September: Carnivorous Plants, Begonia/Gesneriad Show, Dahlia Show.
- October: Ikebana Show, Fall Fest, Halloween Dog Parade, Robots and Plants exhibit.

- November: Night Lights begins.
- December: Night Lights.

Thus, in gardens one might suggest that event staging is dominated by:

- plants in bloom (tulips, daffodils, and daylilies) and invariably held in connection with local and regional plant societies;
- holidays/school vacations;
- weddings;
- seasonality; and
- new/specialty audience recruitment.

In the case of Tower Hill Botanic Garden, the garden claims that, strategically, it does not have an event every week, two weeks, or monthly; it depends on the time of year and there is no set formula. As events drive attendance, there are fewer during the growing season and blooming time and more in the off-season. The average event attracts 800 person per day, but a blockbuster event may bring as many as 1500 persons to the garden.

Events are designed to bring in new audiences, but generally the garden does not necessarily pick an audience and create an event for that audience. Instead, the garden tries to make events that will be fun for everyone and markets different aspects of those events to the people who might be interested in those aspects through targeted social media marketing.

A Typology of Events

Mega-events

In the garden sector, there are very few mega-events. The shows of the International Association of Horticultural Producers (AIHP), often called the World's Fair of Flowers, are held occasionally, with the most recent ones being in Venlo, Netherlands in 2012 and the Taichung World Flora Exposition in Taiwan in 2018/19. Floral World Fairs must be considered a mega-event as, by AIPH category designation, it is a horticultural exhibition with international participation lasting between 3 and 6 months and at least 25 hectares in size.

Even though periodic hallmark events are more common, they are for a shorter period and thus attract fewer tourists. The annual Chelsea Flower Show and the Hampton Court Palace Garden Festival in London,[2] the Canadian Tulip Festival in Ottawa, and Floriade in Canberra are the best-known garden tourism hallmark events. In the case of both mega-events and hallmark events research has been limited to the economic impact of the events, in part because, as Getz and Page (2016b) indicate, they have to "prove their value in economic terms".

In the garden sector, regional and local events are by far the most common, and most botanic gardens will have events every month. Most gardens have festivals and events to drive attendance, with the secondary reason to bring in new visitors and perhaps cater to, or attract new, membership.

As there is such a range of events in the garden world, what follows is a selection of events and special attractions from around the garden world. Examples are chosen from art events, rock concerts, urban events, and the bizarre and different.

Regional events

New York Botanical Garden Orchid Show

The annual NYBG Orchid Show, running from late February to the end of April, has been a major

event in the garden for 17 years. It attracts some 250,000 visitors and is themed yearly to provide a different perspective on the event every year. Thus in 2019 it was Orchid Show; Singapore to celebrate the "city in a garden" and was held in conjunction with Gardens by the Bay, Singapore and the Singapore Botanic Gardens, both of which have a long history of orchid development, exhibition, and growth. In 2019 the orchid show attracted 15,000 visitors and within that number were over 500 who paid US$38.00 to visit the orchid show on one of 11 evenings to view the orchids under lights. As part of the ticket price visitors received entertainment, food, and drinks.

Rock bands at Idaho Botanical Garden, Boise

It is perhaps a unique phenomenon of attractions like gardens, zoos, and museums that, within their visitor numbers, many of the visitors are members of the garden, zoo, or museum. Membership has as its primary benefit reduced admission and preferred access to programs, but gardens with a strong membership base often or invariably must put on events to give added value to the members. Thus, the percentage of visitors to a garden event that are also members would seem valuable for operations and yield management.

In 2018, the member percentage of general admissions to major events at Idaho Botanical Garden was as shown in Table 7.1. Idaho is one of the more ambitious gardens in attracting "big names" to its garden events; Table 7.2 indicates the importance of such big names to the garden.

This example from the Idaho Botanical Garden suggests that members are a key segment of any large event the garden stages, and it might also be hypothesized that for the smaller local events of Getz and Page (2016b), membership numbers would be even higher.

LEGO festival at Reiman Gardens, Iowa

The idea that regional events, and particularly blockbuster shows, in gardens are confined to large gardens is not valid. In 2012, Reiman Gardens, a 17-acre public garden at Iowa State University in Ames, Iowa, decided to create a regional

Table 7.1. Membership attendance at major events, Idaho Botanical Garden, Boise, 2018. Unpublished results courtesy of Idaho Botanical Garden, 2020 and reproduced with permission.

Month in 2018	Number of members attending events[a]	Total number of visitors at events[b]	Percentage of members at events
April	874	2405	36
May	1185	4624	26
June	1013	4149	24
July	809	3416	24
August	637	2646	24

[a] Under the garden's previous membership categories were levels such as "Individual Plus", where the cardholder could bring another person – usually a spouse – into the garden. This guest would not be in the garden if not for the membership, so they are included in the count.
[b] Membership categories included are reciprocal garden memberships and plus levels.

Table 7.2. Attendance and selected big name attractions, Idaho Botanical Garden, Boise, 2017/18. Unpublished results courtesy of Idaho Botanical Garden, 2020 and reproduced with permission.

Band/attraction	Date	Concert attendance	Total admissions	Percentage of total visitors to garden that month
The Avett Brothers	Sept 2018	4000	11,171	36
Steve Miller and Peter Frampton	Aug 2018	3369	18,000	19
ZZ Top	Aug 2018	2710	18,000	15
Carlos Santana	Jun 2017	4153	19,866	21

event as a solution to the high lease prices of exhibits that were out of range of its budget.

The staff utilize a process called "dimensional design" when planning annual themes. This involves all staff brainstorming on how all elements of an exhibit might overarch into displays, programs, events, and activities. The proposed theme for that year was "Connections within Nature" and dozens of ideas were winnowed down to using an extremely popular "connector", highly recognizable by all age groups: the LEGO® brick.

Staff conceived of several sculptures made entirely out of LEGO bricks and discovered there were thirteen LEGO Certified Professionals in the world who were currently using LEGO bricks to create sculptures and art forms. Through Iowa State University, the gardens sent proposals to these artists. Only two were in the USA and only Sean Kenney in Brooklyn, New York was creating sculptures that aligned with the proposed concepts.

When Kenney submitted his bid, it became apparent that the cost of creating an exhibit of this magnitude would still be prohibitive to even doubled attendance and revenue for a garden of this size community. Rather than abandon the concept, staff approached other public gardens to see if they would lease the 16-piece exhibit after its summer run. Three gardens committed to 12-week leases in 2013, which not only paid for construction but also returned some profit.

The exhibit had 13 sculptures ranging from 6 in to nearly 8 ft tall (Fig. 7.3), with the largest being a mother bison with 45,143 bricks. All the sculptures together represented nearly 500,000 bricks.

For Reiman Gardens particularly, the results of this exhibit were outstanding:

- admission dollars increased by 149% over the average of the previous 6 years;
- total paid attendance increased by 78% over the average of the previous 6 years;
- attendance of around 65,000 in 2011 jumped to 105,581 in 2012 with 97,399 from April through October, the months the exhibits were on display;
- gift shop revenue increased by 42% over the average of the previous 3 years;

Fig. 7.3. One of 13 sculptures in the first Nature Connects® exhibit, the 8 ft hummingbird and trumpet flowers, made of 31,565 LEGO® bricks, was a major attraction conceived by and displayed at Reiman Gardens, Iowa State University, Ames, Iowa, in 2012.

- membership revenue increased by 22% compared to the average of the previous 6 years;
- private rentals increased by 20% compared to the average of the previous 6 years; and
- press covered the show on an international level, including in countries such as Belgium, Japan, and China.

Due to the profit potential for a small garden, this exhibit went on to ten more sites until it was retired at the end of 2016. There was enough demand that a second traveling exhibit with 14 pieces was created in 2014 – that exhibit has been in 12 sites[3] and was retired at the end of 2019. Two more exhibits were created in 2015 and have traveled to 18 sites; and a special exhibit featuring extinct and endangered animals was created in 2016 at the request of zoos and has toured six major zoos. The last three exhibits will all retire in 2020.

Eight sites have hosted Nature Connects®: Art with LEGO® Bricks more than once. There is only one case where a site did not increase attendance (Hawaii) but otherwise, on average, doubled attendance was the norm. Because exhibits run a course of popularity and can be repeated only so many times before they lose their excitement, there are no plans to build any more Nature Connects exhibits but a new nature-connecting exhibit is being created for 2019 based on oversized toys and games.

Garden Walk Buffalo

On the Thursday before the last weekend in July 2019, the author was in a garden in Buffalo, New York that was open for Open Days Buffalo, the precursor to Garden Walk Buffalo that was to be held on the Saturday and Sunday of the upcoming weekend. While the author was in the garden, five ladies entered. All were from Seattle, had been in Buffalo since mid-week, were visiting both the Open Days and the Garden Walk, and were to return home the following Monday. They knew of Buffalo's garden walk and open days from following blogs, the website, and other garden awareness outlets and had planned the trip months in advance. They were staying six days in a local hotel occupying three rooms. While they chatted to the house owner, a bus trip from Florida with 48 people arrived to see the garden and the garden owner said this volume and origin had been a common occurrence since Open Days started. This vignette is probably typical of the size and reach that Garden Walk Buffalo has achieved. Started as a one-block community opening in 1995, today Buffalo Garden Walk is the largest free open garden event in the USA with 73 gardens open for Open Days and 456 gardens open for Garden Walk Buffalo. It is estimated that in 2019, 75,000 visitors came to Buffalo Garden Walk (see Fig. 7.4). In the field of garden tourism events, in 2010 Buffalo Garden Walk was clearly poised to grow by quantum leaps. Since 2010 several administrative and program changes have been made that have almost turned it into a mega-event. The changes have been:

- Garden Walk Buffalo has grown from 365 gardens to 435, with clearly an increase in quality of garden.
- The organization in charge of the event changed in 2007 from the local, Elmwood Village Association to Garden Walk Buffalo.
- In 2014 Garden Walk Buffalo was incorporated[4] into Gardens Buffalo Niagara and thus moved from just a weekend garden event to include:
  - a two-day garden art sale, which in 2019 had 42 vendors – local artists and photographers – paying US$75.00 for a table;
  - themed bus tours;
  - a garden bike tour;

Fig. 7.4. Some of the visitors at Garden Walk Buffalo, New York, 2019.

- ○ a front-yard garden competition between landscapers;
- ○ garden art exhibitions;
- ○ garden-themed murals;
- ○ speaking and educational events; and
- ○ a Tour of Open Gardens[5] on Thursdays and Fridays (to extend those weekend visits!), consisting of 70 gardens in 2019, but not free – for a fee of US$10.00 (US$20.00 online or US$5.00 using an app).
- • In addition, the group charged itself with helping to promote ALL the regional neighborhood weekend garden tours.

The garden walk has been so successful that a book, *Buffalo-Style Gardens* (Cunningham and Charlier, 2019), has been produced setting a new garden design category in the horticultural industry and garden design schools.

Of interest, in 2019 the local Convention and Visitor Bureau (CVB), Visit Buffalo Niagara, devoted US$25,000 to a social media campaign to attract existing visitors interested in Buffalo but also new (Millennial) audiences. The results were remarkable. Initially the campaign targeted older women and the lesbian, gay, bisexual, and transgender (LGBT) community in the CVB target markets and the CVB produced a dedicated website for interested parties to respond to. By July 25, the start of the 2019 Garden Walk Buffalo, the campaign had driven 20,230 clicks to the newly revamped gardens content on the CVB website and had 175,872 eyeballs on the garden videos.

In the 30 days of the campaign, the marketing had driven more traffic to the gardens' page than to the homepage (27,099 sessions total thus far). This may be compared to the same period in 2018, when the gardens' page attracted around 800 views.

Finally, Facebook ads automatically optimize and as a result, automatically shift dollars to the ads and demographics that are outperforming the others. The CVB ads have begun to shift towards the LGBT community and the full-length video now has the most views.

For the future, showcasing more gardens is not a priority, but rather to link Buffalo-style gardens into an international tour of gardens starting possibly in Toronto and moving to the Royal Botanical Gardens in Hamilton, Ontario, Canada, then into Buffalo, and ending in Rochester, New York or a similar garden destination.

Tower Hill Botanic Garden's Botanical Tattoo Weekend and "Stempunk" Weekend

Tower Hill Botanic Garden's first Botanic Tattoo Weekend was held July 8–9, 2017 and included live tattooing, an art exhibit, face painting, henna, custom Tower Hill temporary tattoos, and a participatory collaborative art project. The idea was to attract new visitors to Tower Hill. Marketing efforts included a guerrilla campaign to hang stylish post cards in tattoo shops throughout the region, although most of the marketing was driven through event marketing using Facebook. The Facebook event gathered 5300 responses, which at the time was one of the garden's most popular Facebook events ever. Another marketing campaign invited people with plant-inspired tattoos to post them on social media and tag Tower Hill. Tower Hill then held a round robin online vote letting users determine the winner, who won a cash prize. The results of the tattoo weekend were excellent, so much so that it is now an annual event in the calendar (Table 7.3).

The "Stempunk" Weekend was held February 17–18, 2018 and the goal was also to attract new visitors to Tower Hill Botanic Garden. The garden had a celebration of flowers inspired by Steampunk, a literary and artistic genre that mixes elements of modern technology, fantastical fiction, and Victorian-era history and fashion. Visitors experienced imaginative floral arrangements and explored Tower Hill's Steampunk-inspired conservatory displays. Many participated in a botanical Steampunk costume contest or took in a talk on gardening. Notwithstanding February was a cold and uninspiring month, the garden had 996 visitors on the Saturday of that weekend and 753 on the Sunday for a total of 1749 visitors.

Seasonality in garden tourism events: Atlanta Blooms! tulip show

In 2010, Atlanta Botanical Garden found itself with a 4-month period, January to April, with a significant diminution in visitation and yet significant ongoing expenses for maintenance and upkeep of the garden. As a result, in 2011, the botanic garden initiated Atlanta Blooms! – 300,000

Table 7.3. Visitors to the Botanic Tattoo Weekend, Tower Hill Botanic Garden, Worcester, Massachusetts, 2017. Unpublished results courtesy of Tower Hill Botanic Garden, 2020 and reproduced with permission.

Date	Day	Weather	Temperature (°F)	Number of paid admissions	Number of free admissions	Number of members admitted	Total admissions
July 8	Saturday	Rainy	82	534	73	364	971
July 9	Sunday	Sunny	77	689	151	604	1444

Table 7.4. Visitation to Atlanta Blooms!, Atlanta Botanical Garden, Atlanta, Georgia, 2010–2019.

Year	Number of visitors in March and April	Comments
2010	45,000	Started with marketing budget of US$200,000
2011	60,215	26% increase over 2010
2012	53,749	Early warm spring with early blooming and sporadic coverage
2013	41,082	Record rains
2014	47,667	
2015	47,601	
2016	47,233	Budget reduced to US$75,000
2017	51,112	Budget reduced to US$50,000 but harsh winter and rainy spring
2018	60,715	Strategy to reduce marketing and track numbers. No loss of numbers apparent and thus may be a tradition at the garden
2019	79,349	Favorable spring weather

tulips that bloomed in the months of March and April. The result was an attraction that is now part of the product mix of the garden, attracting almost 80,000 visitors out of a total yearly visitation of almost 1 million, and in a period that was historically devoid of visitors and thus revenue. Table 7.4 indicates the growth over the past 8 years.

In 2013, the garden surveyed the Atlanta Blooms! visitor with the aim of establishing the specifics of the visitor to this one event and particularly to understand their motivation and assess whether the event could become a spring tradition at the garden. Significantly, the majority of visitors were female (71%), average age increased to 48 years of age, and 36% of visitors to Atlanta Blooms! were first-time visitors. Of these first timers, the average age was 41.8 years. For the majority of visitors (70%), Atlanta Blooms! was the reason for their coming, with enjoyment of the plants/flowers and family time the major drivers. Of particular interest was the 14% who indicated that Atlanta Blooms! and the specific flowers (tulips, daffodils, orchids) were the way they recognize spring had arrived ahead of Easter or any other flower festival (Dogwoods).[6]

The last area of research was the satisfaction rate and the intent to return. Sixty-nine percent of visitors said Atlanta Blooms! exceeded their expectation and, using a Net Promotor Score quantifying the effect of "raving fans, brand haters, and the indifferent", 86% of Atlanta Blooms! visitors scored the display 9 or 10 out of 10. The visitors "likely to visit Atlanta Blooms! the following year" was over 75%, with some audiences (members) in the 90% range; of the first-time visitors, more than half indicated they would come again in the following 3 months.

Local events

Chelsea Fringe[7] Festival

The Chelsea Flower Show, running for four days in late May, is probably the most iconic flower festival in the world but is limited in time and numbers, catering to only 157,000 visitors in total and for which tickets are much sought after and hard to obtain. Thus it was in 2012 that the garden writer and garden historian, Tim Richardson, thought up the Chelsea Fringe Festival as an alternative to the Chelsea Flower Show.[8] The first year saw 100 events and by 2018 it had grown to 250 separate events planned by 400 applicants with 24 separate categories. It is anticipated that over 200,000 people attend an event sponsored by the Fringe over the nine days the festival runs. The "events" take the form of walks, talks, performances, workshops, dinners, and "on-street happenings". Among the more eclectic and certainly "new" urban garden tourism events, as features of the Fringe, are edible bus stop gardens, guerilla gardens complete with guided tour, vegetable orchestra, and rooftop dinners on green-roof buildings. The venerable Inner Temple grounds in central London host a non-traditional dog show in which competitors compete in five exciting categories: Garden agility; Best floral outfit (dog and handler); Dog with the waggiest tail; Best junior handler (under 12 years); and Public's favorite dog. Most events are free, and all work is volunteer in nature, courtesy of some 400 volunteers.

Media exposure is extensive from television, radio, and broadsheet newspapers, particularly the *Guardian* and *Daily Telegraph*, newspapers that are renowned for their garden coverage. Such was the attraction of the Fringe that in the first year the Duchess of Cornwall spent a whole morning looking at Fringe displays in Hackney and Shoreditch, two of what most people would think are undistinguished London boroughs.

Since 2012, fringe festivals and activities have been staged in other parts of the UK (Aberdeen, Brighton, Bristol, Cambridge, and Kent) and Fringe has now gone international with festivals in Bergamo, Brescia, Florence, and Milan in Italy, as well as Japan, Australia, and Slovenia.

Hollister House workshops

While regional events are the most common in the product and marketing mix of botanic gardens, local events, as defined by Getz and Page (2016b), are very important to small, usually privately owned, gardens. Hollister House Garden is a garden in northwest Connecticut described by its owner as an English garden transported to Connecticut. Thus, while it has significant design elements of an English country garden, it also has elements that make it unique. As visitation is low (and desired to be so by the owner), the garden hosts weekly events through the summer. The focus of the events is art and particularly painting and photography. Not only are artists encouraged to come and produce their own works, but workshops are also provided in a custom-built barn. The choice of topics is informed by a survey in 2015 that asked the 300-person mailing list what they would like to see offered at Hollister House Garden. The results are shown in Table 7.5.

Of additional interest, over one-third of garden visitors come from over 50 miles away to view the garden and the loyal visitors (over 50% have visited Hollister House Garden more than three times) prefer to receive their information by newsletters that are emailed to their computer.

Of drag queens and daylilies: Allen Centennial Garden, Madison, Wisconsin

One of the most unusual local events to take place in the garden tourism world in recent years has been the Drag Queens and Daylilies Program at Allen Centennial Garden at the University of Wisconsin–Madison. The garden was desirous of making plants cool and accessible to a new audience, and thus they contacted Lucy von Cucci and Miss Diana to combine drag queens and horti (culture). Some 300 attendees were able to interact and learn about the world of both drag queens and daylilies. The event started with a lip synch performance by Lucy to Kesha's "We R Who We R", followed by a stand-up comedy show. This led into a discussion about how traits are passed along to different generations, either through drag or plant breeding, and from that a demonstration of artificial pollination was given using the daylily.[9] Attendees then were taught all about the passage of traits in plant breeding and finally got to do their own plant crosses and hybrids using a paint brush. All new hybrids could be named by the attendees. The goal, of course, was to get people thinking about horticulture in a new way and to introduce the topic of plant breeding and genetics to a new audience. Originally presented as a workshop in 2019, it was woven into being part of a *Best. Friday. Ever* series, which typically sees about 300–500 people. The event is primarily advertised through Facebook, along with marketing to a small email list of about 1500 names.

Table 7.5. Survey of level of interest in garden-related events and programs, Hollister House Garden, Connecticut, 2016. Unpublished results courtesy of Hollister House Garden and reproduced with permission.

	Number of respondents (%)				
	Not interested at all	Somewhat interested	Neutral	Interested	Very interested
Horticulture, gardening, plant care, propagating, planting, and pruning	3 (2)	5 (3)	17 (10)	81 (46)	71 (40)
Botany: plant science and education	13 (8)	23 (14)	42 (26)	53 (33)	30 (19)
Garden and landscape design	3 (2)	4 (2)	10 (6)	63 (36)	96 (55)
Family and youth programs	65 (45)	23 (16)	34 (24)	16 (11)	6 (4)
Art in the garden: painting, illustration, photography	20 (13)	22 (14)	31 (20)	47 (30)	38 (24)
Performance in the garden: music and dance	29 (19)	20 (13)	30 (19)	48 (31)	29 (19)
Arts and crafts: arrangements, ikebana, wreaths	30 (19)	18 (11)	43 (27)	41 (26)	26 (16)
Wellness in the garden: yoga, Tai Chi	53 (36)	22 (15)	35 (23)	27 (18)	12 (8)
Guided garden tours	10 (6)	8 (5)	26 (16)	64 (39)	58 (35)
Antiques and decorative arts	16 (10)	15 (9)	24 (15)	63 (39)	43 (27)

Garden events going to the dogs: Fidos After Five

With 89 million dogs in the USA in 36% of the households, dog owners and their dogs represent a significant market for gardens. Of course, the presence of dogs in gardens can be somewhat controversial but a number of botanic gardens have circumvented this problem by opening up the garden on specific days and at dedicated hours to cater to (well-behaved) dogs and their owners (who scoop). Lewis Ginter Botanical Garden in Richmond, Virginia opens the garden to leashed dogs twice per month on Fridays (Fidos on Friday, or Fidos After Five, or even GardenFest for Fidos). The result is as many as 5000 dogs in the garden (with owners) on a summer's evening at a time when the garden might be closed or have lower attendance.

From blooms to booze: using brews and barley for the Millennial visitor

The past 5 years have seen a remarkable growth in craft breweries in the USA.[10] In 2018, there were 7450 craft breweries in the USA – up from the 1514 counted 20 years earlier and still growing at 12.9% per year.

The fact that two of the major ingredients in beer making (hops and barley) are botanicals, as well as the environment and ambiance of (botanic) gardens, has made beer tasting and consumption a significant new event in gardens. Norfolk Botanical Garden, Virginia may have gone the furthest by dedicating a portion of its site to creating an educational garden, the Grains and Hop Garden, and growing both in it. Furthermore, it runs courses on the horticulture and history of beer and gardening for the home brewer, and provides live plants to a local brewery, O'Connor Brewing Co., from which the brewery makes a specialized Norfolk-sourced beer.[11] Of course, beer tasting alongside food trucks and live music has become a major event in the garden with over 300 participants.

The importance of weddings

There are 2.5 million weddings in the USA every year[12] and the reception, including the venue, can be as much as 50% of the wedding costs. This cost is mostly for food, but church and especially hall rental can be significant. Generally, a public garden is considered a less expensive alternative to a church or hall, but US$1000 for use of the garden facilities is not considered anomalous. The attraction of a botanic garden appears to be that couples increasingly desire a unique, beautiful, or unusual outdoor setting, particularly for the ceremony, and a wedding in a garden meets all those criteria. Often the choice of a garden is described by wedding planners as being a "boutique" venue.

Trends in garden usage for weddings are instructive. At Queens Botanical Garden[13] they have noticed that the "product mix" has changed over the years. In the financial year 2008 they had 387 parties at the garden to take pictures either in the Wedding Garden or on the main grounds. The frequency was compared to a revolving door, with a friendly guard posted at the entrance of the Wedding Garden to assure the timely entrance and exit of each wedding party.[14] One of the earliest coordinators insisted that people come in person to sign up, so that she could emphasize that parties needed to be timely. If they were not on time to enter the Wedding Garden they were then escorted to another part of the garden.

As a testament to the unchanging core attraction of a garden being a wedding venue, a banker who only recently found out she had her wedding pictures taken at Queens 30 years ago said, back then, everyone wanted to be at the garden for their pictures! Remarkably, there was a time when using a "botanical garden" for anything but the plant collections and education was frowned upon; even holding public programs, including concerts, was looked upon with disdain. The use of gardens for weddings can be traced to the 1980s when a speech on using gardens as weddings venues was given at an AABGA[15] regional meeting that was held at Longwood Gardens about public programs in botanical gardens. This speech and suggestion were considered revolutionary. Now, of course, almost all public gardens are doing whatever they can to bring more people and build earned revenue!

In 2017 an average of US$2783 was spent on wedding photographs but with the development of cell phones that take pictures, the photo business declined. In addition, staff preference for what to include in the garden offering has also changed. A staff person who comes from the hotel/hospitality sector will focus on wedding receptions and is always seeking to "upsell" the wedding party by offering a cocktail reception at the entrance to the Wedding Garden or some other beautiful place. For garden personnel, receptions are more challenging and require constant monitoring of the music to keep the sound level down and thus the neighbors happy. Also, a garden makes more money with ceremonies, as multiple ones can be booked in a day, whereas the set-up and time for a reception mean the garden can typically book just one per day. Therefore, the trend, especially at Queens Botanical Garden, is headed back in the direction of more ceremonies.

Finally, gardens also get inquiries for memorial services, christenings, and even baptisms.[16]

Of course, in the Northeast USA weddings are very seasonal, so the wedding business only extends from late April through late October. There is a growing interest in mid-week. And then there are weddings that take account of lucky numbers – like 7/7/07, or things to do with the number 8 which is lucky in some Asian cultures. At these times Queens Botanical Garden gets even more calls and bookings!

The accommodation of weddings in gardens makes internal competition for space robust; education wants space for Saturday programs, so do staff from public programs, head gardeners have to worry about the location of compost, etc. … and all events need rain plans. The garden has also found that the weddings function a bit as public programs in and of themselves: with brides and grooms dressed in their "native dress" or sometimes in Halloween costumes and always in beautiful outfits no matter what; people beyond the Wedding Garden guests come to watch the wedding parties! As this was being written, staff at the Queens Botanical Garden were talking about a big reception coming up – 500 people. "Is it an Indian wedding?", the director asked. It is! Indian weddings are typically large – and undeniably beautiful!

Case Study 7.1: Events at Queens Botanical Garden, Flushing, New York

Music in the Garden

Queens Botanical Garden has had a tradition of presenting world music for many years. Following are some of the outdoor concerts that have been held at the garden; this is just a partial list of bands and countries or genres to illustrate what has been done over the last 7 years. Each musical group brings a specific audience to the garden and the garden, in turn, is required to target marketing efforts.

1. Zikrayat, Arabic (Egypt, Lebanon, and the greater Arab world).
2. Funky Dawgz Brass Band, New Orleans.
3. Women's Raga Massive, India.
4. Mar Sala Band, Spain (with touches of rumba and Brazilian rhythm).
5. RIVA & Bohio Music, Haiti.
6. Chai Found Music Workshop, Taiwan.
7. Quintet of the Americas, Western Classical.
8. Samba New York!, Brazil.
9. Maestro Khan, India.
10. NYC Ska Orchestra, Jamaica.
11. NYC Klezmer Band, Eastern European Jewish.
12. The Rough Dozen, American Songbook and Pop.
13. Lara Hope & the Ark-Tones, Rockabilly.
14. Ballet Folklórico de México, Mexico.
15. Retúmba, Puerto Rica (with touches of Cuba, Dominican Republic, and more).
16. The Ebony Hillbillies, Bluegrass.
17. Radio Jarocho, Mexico.
18. Yowana Sari Gamelan Orchestra, Bali.
19. Alicia Svigals, Klezmer.

20. Wei-Yang Andy Lin (erhu) and Nan-Cheng Chen (cello), Taiwan.
21. City-wide Locating the Sacred Festival, produced by the Asian American Arts Alliance, in which traditional artists from Asia and Africa present authentic and inspiring music and dance grounded in spiritual journeys and beliefs:

    a. Mala Desai (traditional Odissi dance), India;
    b. Alhaji Pappa Susso (kora, a 21-stringed harp-lute), Africa;
    c. Doo-Yi Park (kayagum, a zither-like instrument), traditional Korean; and
    d. Ikhlaq Hussain (sitar), classical music of North India.

The garden is getting more inquires now for technology/WiFi, etc. and the garden must answer the question "do we resist such things as part of being at a garden is to 'unplug' or do we not?". Queens Botanical Garden has added some hotspots but "wiring" the whole garden is expensive for any garden (US$40,000 or so). For teacher training (and Queens does a lot!), the garden has found that "access" is more and more needed and so it may be an addition that ultimately becomes mandatory.

Notes

[1] Although the limitations on gatherings by crowds as a result of the coronavirus may make events in gardens less popular or less possible (see Chapter 12).
[2] Both organized and run by the RHS.
[3] Almost entirely in gardens, although some were within zoo grounds.
[4] Along with the National Garden Festival that was in existence at the time.
[5] Modeled on the National Garden Scheme in the UK.
[6] Piedmont Park, adjoining the Atlanta Botanic Garden, has held a Dogwood Festival every April since 1936. Atlanta Blooms! now is a bigger association with the advent of spring than the Dogwood Festival.
[7] "Fringe" because it operates on the fringe of an established event or festival and works on an "open access principle" of if it is interesting, legal, and on the topic of plants, gardens, or landscapes, it can be part of the fringe.
[8] The Chelsea Fringe Festival operates with the blessing of the RHS that stages the Chelsea Flower Show.
[9] The daylily was chosen because, like drag queens, they come in a variety of colors, bloom for only a few hours, then fade and go flaccid.
[10] This book uses the term "craft breweries" to cover all beer producers (including home brewers, micro-breweries, and brewpubs with no minimum production), those who sell beer (often breweries just buy other brewers' product) and that are usually independent and members of the national US Brewers' Association.
[11] On the day this was written the brewery had just contacted the garden for a specialized herb for a specialty beer. The garden had the herb and sent a quantity to the brewery.
[12] With an average cost of US$33,391 (excluding the honeymoon); according to studies from the UK and that are similar to studies on wedding trends in the USA, wedding costs are often significantly higher because of the wedding industry itself. The research found that DJs, photographers, and florists routinely

charge more money for services related to a wedding than those for a similar size function or party (https://en.wikipedia.org/wiki/Wedding_industry_in_the_United_States#cite_note-7, accessed August 4, 2020).

[13] Chuck Wade, a former Director of Queens Botanical Garden (and who still lives very near it), had said that Queens was the first Wedding Garden in the USA … when asked how he knew this he said, "Well, I have never heard anyone else claim otherwise!".

[14] Of course, brides are notoriously late, so he had much to deal with – and always did!

[15] AABGA (American Association of Botanical Gardens and Arboreta) was the forerunner to the APGA (American Public Gardens Association).

[16] In the case of the baptism, a father who died young had a memorial tree planted for him. When his girls were older, the mother and grandmother of the two children to be baptized asked about baptizing the two girls here, in front of their Dad's tree. The wish of course was granted.

Buffalo Garden Walk in Buffalo, New York attracts over 85,000 visitors over one weekend in July. Photo courtesy of Jim Charlier, Charlier Design.

Fido's on Friday permits guests to bring their dogs into the Lewis Ginter Botanic Garden, Richmond, Virginia on special designated event days. Photo courtesy of Lewis Ginter Botanic Garden.

Tower Hill Botanic Garden's one day Daylily Show is one of their most popular events of the Year.
Author's own photo.

Miss Diana shows participants a hybridization technique in Allen Botanic Garden, Madison, Wisconsin.
Photo courtesy of Allen Botanic Garden.

The Tucson Botanical Gardens hosts dancers for *Dia de los Muertos* – the day of the dead – their annual feast with the dearly departed. Photo courtesy of Tucson Botanic Garden.

The San Antonio Botanical Garden's water features help communicate the story of water in Texas. The watering hole uses recycled water and displays harvested water moving to an elevated tank.

Gravetye Manor Gardens, created in 1885, is the garden of the great British garden designer, William Robinson, originator of the English garden style during the Arts and Crafts movement. Now part of an 18-room luxury private hotel, the gardens are a major attraction for the guests and a tour is run daily for non-guests. Author's own photo.

8

Impacts of Botanic Gardens: Economic, Social, Environmental, and Health

_____

Perhaps surprisingly, there is a paucity of published economic impact studies of botanic gardens around the world.[1] Why this is so is not clear. Certainly, the expense of such studies, usually requiring the use of an expensive econometrics model, is a major consideration but also many have pointed out it is axiomatic that botanic gardens should be part of an informed and educated society and thus little economic justification is required. In the era of reduced budgets and increased accountability, this latter reason may no longer be valid and thus it is instrument upon this book to suggest some of the economic benefits of or detriments to gardens. To this should be added the increasing emphasis and demand for gardens to have not only economic impacts but also social and environmental benefits. As a result, this chapter will describe the current state of economic impacts of gardens but also includes discussions of cultural, social, and environmental benefits.

The economic impact of any business entity, like a garden, can be broken down into three areas:

1. *Direct impacts.* This is direct income (from gate receipts, grants, voluntary contributions, etc.). This is often juxtaposed with expenditures on salaries, indirect costs (pensions), other operating costs, and capital expenditures.
2. *Indirect impacts.* This is the spending for an ongoing supply chain. Thus, in gardens, the purchase of 10,000 tulip bulbs for a spring show from a tulip wholesaler would be an indirect impact.
3. *Induced impacts.* These are expenditures by the workers for other services (doctors, dentists) plus jobs supported by the supply chain.

These impacts in total may be termed the *total operating impact.*

However, the direct and indirect income or spending also tends to circulate through a local economy because, for example, those doctors or dentists, as generators of induced impact, may hire a receptionist or a dental assistant, and they in turn spend their wages in the economy. Econometric models account for this impact by use of a *multiplier* which is applied to the total operating impact. In the case of tourism dollars and local businesses, standard multipliers tend to range from 1.5 to 2.0.[2]

Economic Impacts of Botanic Gardens

This section must surely start with a discussion of a 2018 study by Oxford Economics of the economic impact of the horticultural industry in the UK and specifically the section that discusses the impact of tourism visitation on the impact of gardens, as it is the only study of the impact of garden tourism on a national economy. One of the key findings was that spending on ornamental horticulture and landscaping made a contribution of over £24.2 billion in 2017. Of that figure, £2.2 billion could be attributable to overseas tourists visiting gardens and parks – or one-fifth of all tourism spending by overseas tourists. Based on VisitBritain data, in comparison with other activities undertaken by international tourists, visiting gardens was the most popular activity ahead of such others as visiting museums, visiting castles, and visiting religious buildings. To this figure should be added the £690 million spent by domestic tourists, including both overnight domestic travel and day trips. In total, direct[3] visitor spending on parks and gardens contributed £1.2 billion to the nation's gross domestic product (GDP) and was associated with 32,000 jobs and £265 million in tax revenues. As noted, this has been the only national economic impact study of garden tourism. Thus, what follows are examples of regional economic impacts of gardens from around the world.

Dallas Arboretum and Botanical Garden, Dallas, Texas, USA

Dallas Arboretum and Botanical Garden in north Texas receives approximately 700,000 visitors per year. In 2010 the garden embarked upon some major expansions (children's garden, expanded parking) as well as hosting a major art exhibit – Chihuly Glass. An economic impact study was conducted over the subsequent 3-year period (Briesch, 2017). The findings were:

- the total operating impact in 2013 was over US$170 million to the north Texas region;
- construction and expansion provided an additional US$108 million;

- the Chihuly exhibition saw attendance increase by 42% and attendance growth was sustained with a 10% increase over 2011 figures, the year preceding Chihuly; and
- non-local (in-state) visitors increased by 28% and international visitors by 3%.

Phipps Conservatory and Botanical Gardens, Pittsburgh, Pennsylvania, USA

Phipps Conservatory and Botanical Gardens has been recognized as a leader in the USA in sustainable building technology, a nexus of horticultural display and education, and a showcase for horticulture and fine art exhibits. It publishes an economic impact statement for all interested and potential funding partners. The statement suggests:[4]

- In the case of the latter, one art exhibit generated US$12 million in lodging, US$6.8 million in food revenue, and another US$25 million in indirect spending to the region. In that exhibition almost one-third (31%) of the 400,000 visitors came from outside the region.
- The most important part of the Phipps economic impact statement is that the garden and conservatory is considered a key partner with the City of Pittsburgh economic development agency for employee attraction and retention. Furthermore, when companies think about relocating to the region, they know their investment will be multiplied by the presence of the garden not only due to its tangible beauty but also because it is a major source of revenue in the region. This partnership is shown by the desire of many Pittsburgh companies to debut new products and technologies in the garden.

Royal Botanic Garden Edinburgh, Scotland

Unlike most US gardens, the RBGE hosts over 750,000 visitors per year. There are several facets about the garden and its economic impact study of 2016 that are useful in comparison with other gardens (Arcadis Consultants, 2016). The first is that RBGE has free garden entrance.[5] Second, the greater urban area population of over 800,000 is comparable to many US cities of around 1 million inhabitants. Third, the study quantifies volunteer and educational benefits, both facets that are a major part of the operation of gardens; and finally and most importantly, it receives a significant grant (of £8.5 million) from the local urban authority (with a population of 495,000) to offset operating costs. This translates into a per capita subsidy of £31.75 per year.

The RBGE is a non-departmental public body (NDPB) and supported through grant-in-aid by the Scottish Government's Environmental and Forestry Directorate (ENFOR). Like several US, Australian, and New Zealand gardens,[6] there are four gardens, not one, and the grant-in-aid funding of £8.5 million is augmented by self-generated income streams which total just over £2 million. Self-generated income includes research grants, and other capital grants. In addition, RBGE generates earned income from shops, restaurants, and cafés at the four gardens; and obtains income from investments, education courses, admission charges at the regional gardens, consultancy, events, publishing, exhibitions, and other sources. As a registered charity, RBGE also depends on financial support from the public through membership fees and donations.[7] Based on these figures:

- the total economic output is estimated at £23 million;
- the total value added of this output is £13 million;
- the direct visitor impact is £2–3 million;
- the RBGE has over 200 volunteers (or 47 full-time equivalents) whose contribution is estimated at £1.7 million; and
- the educational benefits are £1.5–2.5 million, made up of over 11,000 learners from schools' programs (nursery through secondary school[8]), teachers, higher education institutions, and education outreach to other countries and gardens.

State Botanical Garden of Georgia, Athens, Georgia, USA

This botanic garden may be the garden most comparable to a regional garden for several reasons. The garden is part of the University of Georgia, has free entry, and thus depends on the State of Georgia for most of its operating budget. Furthermore, in 2010 there was a movement by the University, through the State, to discontinue funding, which resulted in a study that produced data like those presented here, namely an estimation of its economic impact but also its social and environmental contribution to the region. Briefly, the garden is 313 acres in size, attracts 300,000 visitors annually, and is "governed" by a Board of Advisors. Its operating budget is US$2.4 million of which 54% comes from the State, the rest from revenue producing activities such as rentals, educational classes and events (19%), fundraising efforts (13%), and grant funding (14%). Based on this operating amount, economic impact was assessed in 2017 at US$29 million and the grant received from the State represents a contribution of US$7.00 per capita based on the population of Athens of 135,000.

Proposed Delaware Botanic Gardens at Pepper Creek, Dagsboro, Delaware, USA

In 2017, a botanic garden was proposed for Pepper Creek in the Delmarva Peninsular of Delaware. The garden would be located on 37 acres and include 1100 ft of tidal flats on the creek. As part of the preliminary assessment of the potential for such a garden, Rockport Analytics undertook an economic and fiscal impacts study of the proposed garden. The study is one of the most detailed and extensive economic base studies ever undertaken on a garden and shows what such a garden would generate. The report indicates significant construction and suggests operational impacts of the garden, once open and established (2021), will be at US$5.9 million for operations and US$3.5 million for ancillary spending by non-local visitors. Significantly, the report also indicates that:

> A botanic garden would contribute certain aesthetic and educational advantages, provide an assist to ongoing economic development efforts to attract and retain new business, raise property and commercial values and contribute to local civic pride. Moreover, the garden would help to strengthen and diversify the cultural and tourism offering of the county... helping to grow

employment. While these benefits are more difficult to quantify, they are both tangible and significant... for the entire region.
> (Rockport Analytics, 2014)

Hamilton Gardens, Hamilton, New Zealand

Perhaps the most extreme of gardens attracting large numbers of visitors to what should be a regional garden and to which many other gardens may wish to aspire is Hamilton Gardens, Hamilton, New Zealand. Situated within a region of 335,800 residents and within 3 hours of a major international airport, it receives over 700,000 first-time visitors per year, numbers on a par with some of the larger gardens in the world, and around 1 million total visitors. Thirty-four full-time and part-time employees, who, of course, contribute by spending in the community, care for the gardens.

Visitors inspired to visit the gardens from outside the region contribute at least NZ$8.9 million to the local economy (independent assessment in 2007). The development of Hamilton Gardens is estimated to bolster the city's economy by NZ$14.5 million per year (Horwath HTL, 2008), based on increased national and international day trippers and visitors who are motivated by the gardens to stay overnight in Hamilton, making it by far the largest outside contributor to the local Hamilton economy. Future visitor expenditure in Hamilton could reach NZ$7.6 million per annum, with a total output of NZ$12.5 million, after the completion of the garden's current expansion. Following stage two expansion, visitors are expected to spend NZ$8.8 million per annum in the city, with a total output of NZ$14.5 million. This is a 63% increase on the garden's economic benefit to the city as evaluated in 2013.

Hamilton Gardens may now be considered a destination garden since it is located outside of the city of Hamilton next to the main highway that runs the length of New Zealand,[9] making it very easy for tourists to drop in. In an earlier (2004) study, an independent research company, International Consultants Ltd,[10] undertook a survey and estimated there were 1.3 million visitors per year, half of them first-time visitors. Half of the garden's visitors were from the local district, a quarter were New Zealanders from outside of the region, and about a quarter came from overseas, mostly from Asia (China, Korea, Malaysia, Japan, and India). Such has been the success of Hamilton Gardens that the International Garden Tourism Network selected it as International Garden of the year in 2013.

Cultural Benefits of Botanic Gardens

As noted in Rockport Analytics' 2014 report on the proposed Delaware Botanic Gardens at Pepper Creek, the cultural benefits of a garden are difficult to quantify. Much has been written about the need for a wide range of other biologic species (Wilson, 1984) for the development of human culture. Indeed, many have suggested that the requirement for floral and other faunal species is part of the human DNA. Thus, the building of new gardens or the existence of a garden in a community creates envy in many other communities who are trying to add cultural value to their community but are without such a cultural treasure. Indeed, botanic gardens are still being built every year, testifying to their value as products in our society. Furthermore, many botanic gardens become both repositories for our culture (see Kew Gardens' archives) as well as venues where artistic and cultural expression may be showcased. Much of the success of botanic gardens in the world today comes from art displays, concerts and other performing arts, and exhibitions. This is certainly an area gardens seem dedicated to pursuing and developing even further, thus making botanic gardens even more relevant to the cultural attractions of their host community (see Chapter 4).

Case Study 8.1: Cultural Benefits at Glenstone

Throughout the book thus far an ongoing theme is the close integration, some might say symbiosis, between art and gardens. As detailed in Chapter 2, many gardens now feature art installations and many have new directions to bring various art forms – glass, light, sculpture – into the garden environment. Many of the events and attractions embarked upon by gardens and outlined in Chapter 7 feature art installations

and artists performing in gardens, but if there is one place that is the apogee of the melding of gardens and art it is the new Glenstone Museum some 15 miles from Washington, DC in the USA. Started in 2006 and opened in its expanded form in 2018, with over 300 acres of grounds, the museum states as its core experience:[11]

> [We] envision Glenstone not only as a place, but a state of mind created by the energy of architecture, the power of art, and the restorative qualities of nature. At the core of the museum is a collection of post-World War II art, a very personal project driven by the pursuit of iconic works that have changed the way we think about the art of our time.

In the context of garden tourism, the architecture of the museum is designed specifically to integrate into the landscape[12] and even within the main architectural structure an 18,000 ft^2 water court has changing plant life throughout the seasons. The landscape is a mixture of open areas, reforested groves dedicated to native trees, native meadows, and managed watercourses. As a totality the landscape is managed on a fully sustainable basis (see Chapter 9) and the extensive outdoor art even includes a 36 ft floral structure, *Split-Rocker*, by Jeff Koons.[13]

Social Benefits of Botanic Gardens

Again, much like economic impacts, the social impacts of botanic gardens have been little researched and established. Botanic Gardens Conservation International (BGCI) funded and piloted a study in 2011 that asked, "How socially relevant are botanic gardens?" The study (Dodd and Jones, 2010) came up with four significant findings.[14] These are:

- botanic gardens are well placed to educate the public on conservation issues and the human role in environmental change;
- botanic gardens are enhancing their relevance by broadening their audiences and undertaking such projects as community gardening, greening communities, and education;
- botanic gardens around the world are taking action by demonstrating their social worth; and
- future developments in gardens – such as redefining their mission, active consideration

of their social roles, communicating, and advocating for the environment and modeling themselves on places like the Eden Project, Cornwall, UK – have, at their core, social responsibility and relevance.

In Australia, the Australian Bureau of Statistics (ABS) has been a world leader in the definition of what constitutes social capital[15] and its social capital framework breaks down what is commonly called social or civic participation into three types of participation: (i) social participation; (ii) civic participation; and (iii) community support. Australian botanic gardens in general support heavily these three types:

- Social participation is participation in inherently enjoyable activities, valued in their own right; either formal, provided by organized groups, or informal, with family and friends. Membership in a garden as part of its "Friends of the Garden" category is a striking example of this (see Case Study 4.2).
- Civic participation is participation in governance and citizenship including political activities.
- Community support is participation in those activities that are aimed at providing assistance to other individuals, groups, and the wider community that are not directly related to political participation or participation in governance. In this regard, the presence of volunteers in the garden both to assist visitors and to work in the garden would be considered major contributions to social capital.[16]

The ABS then goes on to delineate 19 types of organization that provide social capital to a community, for which the presence of a botanic garden meets many of the organizational needs, namely:

- Recreational groups – gardens provide dedicated space for bushwalking and birdwatching.
- Arts, culture, and education groups – almost all gardens provide space for art exhibitions; and in and of itself, a garden is an *ex situ* museum for the collection of plants, thus meeting the criteria for culture and education.
- Environmental welfare groups – gardens are major contributors to biome conservation and protection.

- School-related groups – all gardens have active child/school educational programs.
- Self-development groups – gardens have active adult education/invited speaker programs.

In total, and compared to other institutions and facilities, the ability of any garden to meet five requirements for recognition as possessing significant social capital must be seen as remarkable, not only in an Australian context but in a world context.

What the community gets out of a botanic garden as social capital

As a specific example of how important the social capital from a garden is, one might again draw upon the comparative example of Hamilton Gardens in New Zealand.

There has been outstanding community support of Hamilton Gardens and this is reflected in NZ$6 million in sponsorship and consistently high satisfaction scores in the city council's annual residents' survey reports. Specifically, Hamilton Gardens meets the following community needs:

- increased economic benefits and tourist activity bring revenue to the city and provide more job opportunities;
- more Hamiltonians are connected socially and culturally through an extended range of activities;
- helps provide a package of world-class tourism offerings in the Waikato (with Hobbiton and Waitomo Caves);
- unique source of city pride and identity;
- creation of additional performance spaces, a destination for educational activities;
- provides for community well-being as a place for relaxation, rest, play, recreation, and enjoyment; and
- enhances the public use of the gardens.

Hamilton Gardens already hosts hundreds of community and ethnic events and family celebrations every year, including:

- The Hamilton Gardens Arts Festival – over 100,000 Hamiltonians and out-of-town visitors to the garden in 2014.

- Wide variety of groups and individuals (inside and outside venues) – music and cultural events, festivals, orienteering, fairs, garden tours, school and club meetings, reunions, family occasions including birthday parties, funerals, and graduations.
- Hundreds of weddings – it is a popular venue for these.
- National rose trials held in the Rogers Rose Garden – presented with the Garden of Excellence Award by the World Federation of Rose Societies in 2006, an award that recognizes exceptional international rose gardens.
- Waikato Institutes of Technology Horticultural School, located in the center of Hamilton Gardens, offers courses in arboriculture, horticulture, floristry, landscape design, and horticultural technology.

Application of social capital in US botanic gardens

In the USA, much like Europe and Australasia, the need for social inclusion has become a defining direction in sociology and particularly the need for social justice.[17] Seven US gardens have embarked on a new direction with a program of social inclusion that is expected to grow significantly in the coming years. One garden, the San Antonio Botanical Garden, Texas, has joined Museums for All, a signature access program of the Institute of Museum and Library Services (IMLS), administered by the Association of Children's Museums (ACM), to encourage people of all backgrounds to visit museums regularly and build lifelong museum-going habits.

The program supports those receiving food assistance benefits via the Supplemental Nutrition Assistance Program (SNAP), whereby if recipients visit the San Antonio Botanical Garden, they qualify for a reduced admission fee of US$3 per person, for up to four people, with the presentation of a SNAP electronic benefits transfer card. Museums for All is part of the San Antonio Botanical Garden's broad commitment to seek, include, and welcome all. According to Melinda Cerda, interim director:

> The Botanical Garden is proud to join Museums for All and provide a pathway that alleviates the financial barrier for our working families. We strongly believe that every member of our community deserves the opportunity to connect to the natural world and experience all the Botanical Garden has to offer.

The Museums for All admission includes access to the botanic garden's 38 acres of garden space, including the Family Adventure Garden with 15 fun spaces that encourage unstructured play and exploration. Visitors enjoy colorful floral displays, true to Texas native areas, and futuristic glass pyramids filled with exotic plants from around the world. As a living plant museum, the San Antonio Botanical Garden is dedicated to the collection, cultivation, preservation, and display of a wide range of plants.

Health Benefits of Botanic Gardens; The Community Social Dimension

While the health benefits of a garden were intensively referenced in the earlier edition, since 2013, the role of a botanic garden in overall community health has begun to emerge. In Boise, Idaho, the botanic garden managers are looking to the Boise Public Library as a model for assessment. They are conducting health impact assessments and look to dimensions of health including physical health, social health, economic health, environmental health, intellectual health, emotional health, and spiritual health. The library writes:

> addressing health should be a holistic endeavor focused on the whole person and the whole community. Health is made up of many interconnected components that must all be achieved individually in order to obtain overall health.
> (Vitruvian Planning, 2018)

The study particularly addressed overweight residents,[18] low-income residents usually without health insurance, their neighborhoods, and residents in need of mental health services. Suggestions for addressing this need were not confined to physical health. Mental health was identified as a problem[19] and to help alleviate this problem access to the Boise River and its green belt was suggested. The fact that the garden is only a short distance from both the river and the library suggests that the Idaho Botanical Garden can be part of the solution and even learn from this approach.

Environmental Benefits of Botanic Gardens

Botanists have identified more than 390,000 species of plants worldwide.[20] However:

- approximately 34,000 are threatened at present;
- two-thirds of the world's plant species are in danger of extinction during the 21st century; and
- of the 20,000 known plant species in the USA, more than 200 had already vanished by the end of the 20th century and another 600–700 species are in imminent jeopardy.

These plant species are in jeopardy because of a burgeoning human population that then affects proximate causes such as deforestation, habitat loss, the spread of invasive species, and agricultural expansion. In total, as a society we stand to lose thousands of plant species worldwide in the next few decades unless there is a concerted and collaborative effort to conserve them.[21] In this regard, botanic gardens provide six major benefits to the world of plant conservation (Wyse Jackson and Sutherland, 2000). They are:

- Horticulture and cultivation skills allow us to grow plants that might be lost in nature, which means their plants' diversity can be conserved in the gardens, but also allows us to consider restoration and rehabilitation of degraded habitats.
- Living collections of plants collect species under various groupings, to maintain a living store of genetic diversity that can support many activities in conservation and research.
- Seed banks and collections of living plants allow species to be safeguarded. Plants must be carefully collected and stored to ensure maximum genetic diversity is retained, and much research is required to determine the best way of storing each species. This is the conservation of plant diversity *in situ*, and botanic gardens are key to this strategy's capacity and success.
- Research and development into plant taxonomy and genetics, photochemistry, useful properties, informing selection of plants that can withstand degraded and changing environments (especially important in face of the threats posed by climate change).

- Education is a strength of botanic gardens that allows them to communicate the importance of conserving plants, reaching out to diverse audiences, and to communicate how this may be achieved.
- Linking plants to the well-being of people and helping conserve indigenous and local knowledge to encourage the sustainable use of plant resources for the benefit of all, as part of sustainable development.

As part of the mission of all botanic gardens, environmental considerations are foremost in the operation of the garden, thus it is difficult to isolate or provide as an exemplar one particular garden for the work it is doing in environmental conservation. That said, the oldest public garden in America, Missouri Botanical Garden in St. Louis, might be the most typical garden to indicate its environmental outreach and is thus featured in Case Study 8.3 for its work in Africa. However, as part of a larger outreach, Missouri Botanical Garden has operations in 35 countries around the world doing research and field work, providing expertise, assistance with fundraising, and communications with the worldwide scientific community. Studies concentrate on the plants of Central and South America, sub-Saharan Africa, Madagascar, China, and North America – in the case of North America, it is the home of the 31-volume *Flora of America* that is compiling a listing of all plants in North America.

Case Study 8.2: Economic Impact Assessment of the Australian Arid Lands Botanic Garden, Post Augusta, South Australia, March 2014

In March 2014, Port Augusta City Council commissioned an economic development report on the Australian Arid Lands Botanic Garden (AALBG) (Benfield, 2014). The findings would appear to be a basis upon which the value of the garden could be ascertained and thus merits inclusion herein.

Essentially the study found:

- direct impact (or gross production value or output) of AU$580,000;
- an "indirect impact"[22] of AU$157,000 and ten more full-time jobs;

- the total impacts are listed as AU$404,000; and
- there is no application of an expenditure or employment multiplier.

The study also refers to other economic benefits of the garden. These are:

- Tourism – the study indicates that 35,000 visitors come to AALBG from Port Augusta and another 30,000 from outside the region; there is no economic value ascribed to either visitor type or any standard tourism multiplier applied.
- Volunteers – the study indicates there are 13 full-time equivalent volunteers at the garden. No source of this equivalency is provided nor is the value, as was done in the RBGE study noted above.
- Social, cultural,[23] and environmental – these values are called (rightly) "non-use" benefits and include:
  o social capital;
  o option;
  o bequest; and
  o existence values.

The study proceeds to explain what these non-use benefits are but does not ascribe either any qualitative or quantitative value to these benefits, suggesting merely that they are present in this garden.

The study, as suggested above, has some serious limitations, some of which are now addressed:

1. The direct impact (from staff and wages/salary) derived from empirical data from the AALBG and as such is considered accurate.

2. The economic impact of the visitors can be estimated from both AALBG data and official Tourism SA data:

   a. AALBG estimates international visitation as 2000 persons but this is an estimate based on departing guest comments, not empirical data. Official figures from Tourism SA, using the data from the Flinders Ranges and Outback Tourist Region, indicate the region gets 35,000 international visitors per year (see above). On this basis and using the AALBG numbers, the AALBG would only attract 5% of those visitors. Intuitively, this is just not believable. Given the garden's location, and the fact that there are no intervening opportunities available once a person heads out to Alice Springs or is coming south from Alice

Springs on the Stuart Highway, one would expect a significant number would stop at the garden. Assuming that, instead of 5%, the garden attracted one-third (33%) of those 35,000 Flinders Range visitors, one might estimate an international visitation of 12,000 visitors. The economic impact of these visitors might now be calculated.

   b. Assuming 12% of the total international and inter-state regional room-nights (274,000 + 391,000 = 665,000 × 12% = 79,000 room-nights) at AU$100.00 per night (Comfort Inn, Port Augusta, average room rate on day of writing), the total economic impact from visitors is AU$7,980,000. However, one must assume that 50% use campers/motorhomes,[24] so the total direct economic impact of visitors (NOT the garden itself) is about AU$4 million. That is the direct impact of visitors staying in Port Augusta and seeing the garden. Now, it is standard practice to add a tourism multiplier of 1.5[25] for the indirect impact of the tourists and one gets a total economic impact of AU$6 million for tourism visitation alone.

   c. These, of course, are gross numbers but justifiable in the context of official Tourism SA data and comparable to visitation numbers and their impact at comparable gardens around the world (see Hamilton Gardens or State Botanical Garden of Georgia above).

3. The third area that might be addressed is those issues of a social, cultural, and environmental nature mentioned above but not assessed in the economic development report.

Arid and semi-arid lands cover one-third of the Earth's land surface and support more than one-fifth of the world's population. Over 60 nations have significant dry land areas with over 30 having more than 75% of their land area classified as desert. While generally devoid of settlement they have been occupied and utilized by humans for tens of thousands of years. Notwithstanding their dry and apparently barren nature, some dry lands have a rich diversity of plants and animals. One North American biome has more than 3000 plants species, many thousands of invertebrates (1200 bee species alone!), and more than 550 vertebrate species. Remarkably there exist only two desert botanic gardens in America[26] and with the AALBG, these three gardens constitute the only botanic gardens that exist in arid zones of the world.

The importance of botanic gardens showcasing arid lands' flora is not only understated in terms of current developments, but arid lands serve to address another major development in the world's dry land regions, namely desertification. Desertification is the degradation of land in dry regions through various factors, including climatic variations and human activities. It affects about 8% of the world's land surface and about 15% of the world's population. Measures to combat it are the focus of a major international effort, the United Nations Convention to Combat Desertification (UNCCD), in which action programs have been drawn up, especially for developing countries in Africa experiencing serious drought and/or desertification. Thus, AALBG can contribute to yet another regional and international issue.

Case Study 8.3: Missouri Botanical Garden and the Study of African Plants

The Missouri Botanical Garden was recognized as a center for the study of African plants as early as 1971. Today Missouri's focus is five major programs:

1. Conserving African orchids. Owing to pressure from land clearance for mining, agriculture, and logging, orchids – which, being epiphytes, rely on other plants for their existence – are severely threatened. Missouri pioneered the use of shade houses to conserve orchids threatened by land clearance such that, today, over 25,000 orchids from 500 species have been grown in these shade houses and at least 50 previously unknown orchids have been identified as part of this program.

2. Plant diversity. Over 170,000 collections have been made from 25,000 species (5% of which were entirely new to science) since 1983, of which 125,000 came from just four countries: Cameroon, Gabon, South Africa, and Tanzania. More recently new initiatives in Guinea (Conakry), Liberia, and São Tomé and Príncipe have given these countries the highest density collection of species per hectare in Africa.

3. The Lower Ogooué delta region in Gabon is being surveyed for its unique floral collections (33 new plants have already been identified) such that management plans for the delta's conservation and management for visitors can be made.

4. Conservation assessments. The Global Strategy for Plant Conservation calls for an assessment of all known plants by 2020. Missouri Botanical Garden has led the way by compiling records for nearly 20,000 tropical African species and showing that probably 35% of these species are threatened. Missouri Botanical Garden also helped establish the regional East African Red List of Threatened Plants to assess their plant life which comprises almost 20% of the region's total flora; and of those plants, 60% may be considered threatened. Such a situation calls for conservation actions that in many cases involve conservation area designation and management.

5. Africa is considered a biodiversity hotspot owing to the fact it is one of the most important regions for biodiversity in the world. As a result, the identification of plants that might then be afforded protection in designated protected areas becomes imperative. South Africa has led the way with its network of botanic gardens in nine distinct biomes, while Missouri Botanical Garden works throughout the African continent to promote national land-use planning by governments, non-governmental organizations, private industry, and other stakeholders to ensure the conservation of these unique plant species.

Notes

[1] In North America and Western Europe, many economic studies are done to support new fixed asset investment as part of a pro forma statement of revenue for financing purposes and requirements. Other studies have been produced to estimate return on (grant) investment for other garden projects. Few studies examine the garden without such major capital investments or a limited-time special exhibit.

[2] In the case of the economy of Port Augusta, South Australia, which is described below, such an economy is often termed an "island economy" and would suggest that the multipliers will be significantly higher than in other regions.

[3] Adding indirect and induced contributions suggests gardens add £24.2 billion in GDP and £5.4 billion in revenues to the government.

[4] Available at: https://drive.google.com/file/d/0B9NBG-omRzfRQzA5Sjl6UW9BN3M/view (accessed July 30, 2020).

[5] Not unlike many gardens in the UK and USA.

[6] Such as Denver, Adelaide, and Wellington.

[7] For a detailed breakdown of RBGE income, see https://www.rbge.org.uk/media/4559/how-we-are-financed.pdf (accessed July 30, 2020).

[8] In the USA, nursery through to secondary school is equivalent to Grades K–12.

[9] In this regard, proximity to interstate highways in the USA, motorways in the UK, or major highways in Australia becomes a significant factor in attracting visitors.

[10] The original study in 2004 was updated in 2008 (Horwath HTL, 2008).

[11] Available at: https://www.glenstone.org/about/mission/ (accessed August 4, 2020).

[12] Which in 2020 won the American Institute of Architects' award for best contemporary building.

[13] Available at: https://www.glenstone.org/search/Jeff+Koons (accessed August 4, 2020).

[14] The study came up with 18 findings, many of which relate to plant collections and are not relevant here.

[15] The World Bank (1999) defines *social capital* as "the institutions, relationships, and norms that shape the quality and quantity of a society's social interactions. Increasing evidence shows that social cohesion is critical for societies to prosper economically and for development to be sustainable. Social capital is not just the sum of the institutions which underpin a society – it is the glue that holds them together" (Smith, 2000–2009).

[16] RBGE values volunteers at £8500 per person, or one volunteer saves the garden £8500 in full-time salary.

[17] See Pollan (1991).

[18] Twenty-eight percent of Boise residents are considered obese.

[19] The study found that 16% of Boiseans have experienced at least one week of bad mental health out of every month – a rate above the state average.

[20] See https://stateoftheworldsplants.org/2016/ (accessed July 24, 2020). The report suggests many more are still to be discovered.

[21] For example, with a significant proportion of arid land plant species being threatened with extinction in the wild, the living collections in botanic gardens have huge potential for *ex situ* conservation and restoration of natural populations. In the case of arid lands species, only 68% of threatened cacti are recorded in botanic garden collections.

[22] Which the study calls "induced impact", failing to distinguish between indirect impacts (money paid to suppliers) and induced impacts (spending by staff and supplier employees); see above.

[23] The AALBG in Port Augusta hosts an annual "Arid Sculpture Exhibition" that runs for a month and showcases the work of sculptors from primarily the northern part of the state, offering a venue that is most suitable and probably not available in other parts of the state.

[24] Who also contribute to expenditures in Port Augusta.

[25] See Steven Wanhill's work (1994) on tourism multipliers.

[26] They are the Desert Botanical Garden in Phoenix and the Arizona-Sonora Desert Museum in Tucson. Furthermore, their collections are dedicated to North American species and their North American collections are unique in the world.

Weddings are a major source of earned income for Gardens. This wedding venue is Mount Coot-tha Botanic Garden, Brisbane, Australia. Author's own photo.

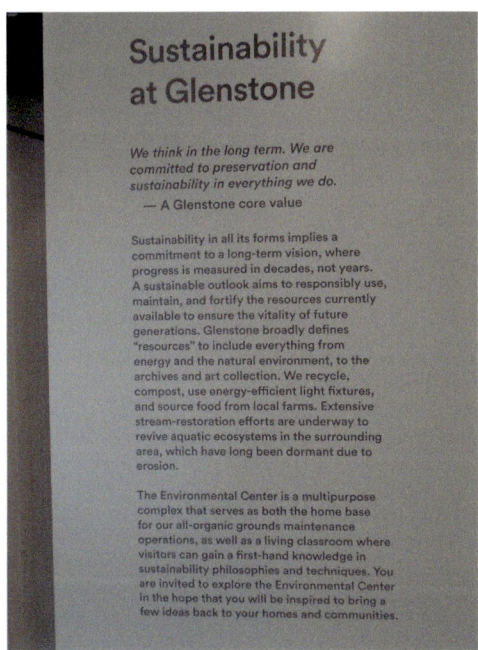

Glenstone Art Museum also promotes sustainability in response to a garden's environmental impacts at the Environmental Center, which is a major part of the garden visitor experience. Author's own photo.

At the Quito Botanic Garden, Ecuador, research on plants and their environmental impacts and issues is conducted in conjunction with Missouri Botanic Garden, a continent away. Author's own photo.

Waimea Valley Botanic Garden, Oahu, Hawaii, stresses the cultural history, impacts and significance of native Hawaii plants and their uses to indigenous Hawaiian people.

The Outlaws and Carlos Santana perform in Boise Botanic Garden. Such a concert may attract over 4,000 people and provide almost 20% of earned income for the month. Photo courtesy of Boise Botanic Garden.

Large Galas held in gardens as part of fundraising were a major source of revenue for many gardens pre-Covid. This gala was held at San Antonio Botanic Garden prior to Covid.

The Garden Museum was a small museum in a church attached to Lambeth Palace, London until renovated and redesigned in 2017. It is the site of the graves of the famous plant collector, John Tradescant, and William Bligh of "Mutiny on the *Bounty*" fame. More recently, skeletal remains of at least 30 burials were found under the floor. Author's own photo.

9

Urban Garden Tourism

_____

© Richard W. Benfield 2021. *New Directions in Garden Tourism* (R. Benfield)
DOI: 10.1079/9781789241761.0009

The world is becoming more and more urbanized. In 2017, 55% of the world's population resided in urban areas with over 80% in the developed world and 33% in the least developed countries living in cities. In the USA 83% of the population lives in urban areas, 75% in the EU, and 85% in Australia. This statistic obscures the fact that in some developing countries, urban populations and densities are among the highest in the world and as a result gardens in these cities have pressure placed on them by surrounding urban growth and development. Thus, in Bogor, Indonesia, site of one of the most important gardens in the world, urban pressure is exerting significant strains on the ability of the garden to remain a green oasis. In Calcutta, India, the Hooghly Botanic Garden is surrounded by urban development and even in a city like New York, Brooklyn Botanic Garden is under threat from the prospect of two 39-story towers being built 150 ft from the garden and thus shading the garden for much of the day. Much of the literature on gardens and urbanization has been concerned with the provision of green space for the residents. Thus urban land-use studies have compared cities with large areas of parks; USSR cities with over 250,000 inhabitants had, for example, 20.3% of their land area under parks – significantly more than Western cities' average of 9.3% – owing to the emphasis put on outdoor space and parks by the former communist governments (Reiner and Wilson, 1979). This green space, of which gardens are a part, has for many years been categorized as "Parks and Gardens" in land-use categories. This classification is still highly prevalent. The steady progression of urban growth has meant, for formal gardens like botanic gardens, that while once they were located on the periphery of cities where land was relatively inexpensive and available, today they have been engulfed by the wave of urban structures that increasingly moved out from the city core. Thus in the USA, gardens in cities like Denver and New York, which in 1959 and 1890 in the case of NYBG, respectively, were in the suburbs, are now surrounded by urban structures and confronting urban issues such as gentrification, parking, neighborhood outreach, and even social issues of homelessness and social integration, equity, and inclusion. Thus, new directions in garden tourism in urban areas are marked by reaching out to new audiences (where neighborhood demography is changing), community programming, local participation in decision making, attracting new ethnic audiences, and overall outreach to their own particular neighbors.

To chart each garden initiative in the areas of urban policy would be exhausting and would not be able to encapsulate all the different initiatives being pursued. Thus, this chapter will highlight a number of urban programs related to gardens that rank as the most noteworthy of gardens' new directions in an urban milieu. The discussion will start with three examples from three of the largest urban conurbations in the world:

1. New York City and particularly Brooklyn Botanic Garden's Community Greening Program.
2. The City of Westminster, London and its responsibilities for over 54 open-space properties and outdoor structures.
3. Philadelphia, Pennsylvania's City Harvest initiative.

The Urban Gardens of New York

New York City may constitute the largest urban area concerned with urban gardens. There are numerous organizations greening the city among which are:

- The New York Restoration Project that transforms unused community space in a neighborhood into community vegetable gardens and beautification display gardens. Started in 1996 by the singer Bette Midler, the goal is to serve under-resourced neighborhoods in all five boroughs and as such is the broadest in coverage. Today many of these community gardens serve new ethnic immigrants by growing indigenous plants locally that are sold to the new residents, particularly refugees, to build community through food. The New Roots Community Farm in The Bronx is one such community garden, providing vegetables for US$1.50 to primarily new resident refugees in the borough.
- Trees New York, which in 2018 planted 220,000 trees in streets, gardens, and cemeteries and provides educational opportunities for neighborhoods and particularly youth on the environment and natural resources.

This initiative is derived from the Million-TreesNYC initiative started under Mayor Bloomberg and that now boasts 675,000 trees in New York City. It is not coincidental that the 2018 annual environmental career day which is part of the Trees New York initiative was held in Queens Botanical Garden. In 2019 the focus was on The Bronx with the planting of 234 trees on sidewalks, parking lots, playgrounds, and any area in need of shade.

- Other community organizations focus on river restoration with, in the case of the Saw Mill River, "daylighting" the river to remove its former underground portions and expose the water with associated riverside walks, reintroduction of species that depend on the river, and rehabilitation of any former polluting sites often designated "superfund sites".

Brooklyn Botanic Garden Community Greening

In 1993 Brooklyn Botanic Garden introduced a community horticulture and urban greening outreach program called GreenBridge based on the view that, at that time, there was an implicit disconnect between the garden – with its railings, walls, and access to a paying public – and the community of which it was a part and which surrounded it on all sides. With this program the garden turned from being an inward-looking attraction to looking outward toward the borough of Brooklyn in a real programmatic sense. By 2006 the program had evolved into five distinct programs and a 2006 strategic planning process was instituted to guide the program into the future. By 2010 the (new) program got underway wherein there are still five distinct programs designed to extend the garden's resources throughout the neighborhood and encourage greater community participation in the programs run by the garden. The five programs are:

- Greenest Block in Brooklyn whereby over 200 city blocks (it has been as many as 240 blocks) compete to be the greenest block in the borough. Judging takes place in the first week of August and the block receives extensive media coverage, prizes for the best block, and almost certainly real estate appreciation and recognition for their efforts.

- Making Brooklyn Bloom is an early spring symposium where individuals can attend lectures, workshops, and exhibits with the goal of making community horticulture more sustainable. This symposium usually attracts over 1000 participants.

- The Street Tree Stewards program had been in existence since the founding of GreenBridge but received a major boost with Mayor Bloomberg's MillionTreesNYC campaign between 2007 and its goal of a million trees planted by 2017. Perhaps the most important outcome of this program was that residents' psychology of urban arboriculture changed from a "not my tree" mentality – believing trees on the streets were the city's responsibility – to residents taking ownership of their trees. Participants receive free tools, training, and permits to prune their neighborhood trees.

- Brooklyn Urban Gardener (BUG) which is a certification program started in 2010 that takes 15 graduate trainees per year and trains them in urban gardening. Certified BUGs then continue to support greening projects at schools, senior centers, block associations, community gardens, and other organizations such that residents and workers in the borough can grow, in an urban setting, their own food. To date, 130 graduates have passed through this program.

- Community Garden Alliance is a network of between 60 and 120 member gardens whose focus is on promoting sustainable gardening practices to support healthy communities of people, plants, and wildlife. It facilitates skill sharing between gardens through workshops and seasonal gatherings and offers an avenue for communication and learning where personal connections are forged, and technical horticultural assistance may be obtained. There is no cost to participate.

City of Westminster Gardens

The City of Westminster, London's central borough, has 116 gardens (some of which are community gardens), parks, and cemeteries, and is responsible for their upkeep and regulation of use for various groups. In addition, within the borough are four gardens administered by The Royal Parks. They are Hyde Park, Green Park, St. James Park, and Kensington Gardens. All of the parks are used extensively by both domestic and international visitors. In total, The Royal Parks estimates that 77 million people use its parks annually. As well as the more sedentary pursuits of walking and picnicking, the parks are also used for major events. The Royal Parks are used for events such as the London Marathon, the summer concerts in Hyde Park, the Frieze Art Fair, and Winter Wonderland, and they are also, like the National Trust properties, vehicles and venues for delivering commercial filming income.[1] However, a number of critics suggest this is a form of neocolonialism that is changing the nature of London's parks in unacceptable ways (see Smith, 2019). In response, The Royal Parks believes that such events add a new dimension to a park visit, attract new audiences to the parks, and generate significant income for both The Royal Parks charity and the wider economy.

A major feature of London's gardens is the certified quality of the green space experience. All gardens in the City of Westminster are candidates for the Green Flag award that judges gardens on eight major criteria covering everything from access through to sustainability and marketing. In 2018, 31 of Westminster's gardens were awarded a Green Flag award.

Philadelphia Urban Initiative: City Harvest

Philadelphia may have the oldest community gardening project providing food for the urban poor in the USA, which was started in 2006 by the Philadelphia Horticultural Society to provide food for the 22% of Philadelphia's urban population, mostly children, who are experiencing food insecurity. The program started with a cooperative agreement with the Pennsylvania prison service that provided extra produce for the outside population; today the prison facilities are an integral part of the growing and supply of horticultural products, particularly seedlings, for the City Harvest program. While the prison supplies seedlings from its nursery as part of the supply chain, the majority of the efforts emanate at Bartram's Garden, America's oldest private botanic garden, where City Harvest has 5 acres of garden plots and horticultural nurseries from which its supplies 150 gardens in the city with 250,000 free seedlings, a tool library, and workshops to teach how to grow the food. In addition the Awbury Arboretum outside Philadelphia provides trees for the Philadelphia Orchard Project, develops a teen leadership corps, and is the location of the Philly Goat Project, a project bringing the arboretum into the community by means of goats therapy, goat grazing (of invasive plants), and goat walks through the arboretum.

The most recent challenge for the City Harvest program is to cater to new immigrants and refugees into the city. Thus, programs like Growing Together and Growing Home reflect the need for foods from these homelands of the new arrivals. While such foods as chili peppers and white eggplants have been staple provisions since the program's beginning, the challenge now is to integrate foods from Southeast Asia, Nepal, and Africa. In one community garden in south Philadelphia the community garden has to cater to five different language groups!

In total the City Harvest program caters to 150 gardens varying in size from small parcels to 2-acre plots that in total supply over 600 food share outlets in the Philadelphia area.

Neoliberalization of Parks and Gardens in Urban Areas

Perhaps the one issue that defines urban parks and gardens over the last 10 years has been the decline of public funding for parks and gardens and the increasing turn towards commercialization of the parks to make up for that lack of funding. Smith (2019) indicates that "92% of UK park managers reported reductions in their budgets in 2013–2015 with 33% experiencing cuts of over 20% in the same period (Smith quoting UK Heritage Lottery Fund, 2016). As a result park managers have been forced to find new sources of revenue and much of that has come from commercial income streams in the form of provision of commercial facilities in the gardens[2], installation of visitor attractions[3], sponsorship, the requirement for permits and commercial licenses, and staging commercial events, the latter causing some observers to consider urban event tourism the defining characteristic of the modern city. This change has been linked to what has been called the "neoliberalization" of urban management and operation or the justification for private-sector involvement in what has historically been a free, government-funded operation including social services and leisure provision. In the USA Krinsky and Simonet (2011) have traced the same kind of change since the crises in US cities in the 1970s and closer to the example in the previous edition of this book, Loughran (2014) examined the way the High Line in New York is "structured and controlled – with commerce and consumption prioritized in the design and regulation of the park" and concluded that the High Line was "an archetypal urban park of the neoliberal era" whereby parks "are increasingly justified, created and maintained via the value they add to the surrounding real estate" (Smith quoting Millington, 2015).

Smith (2019) devotes a whole chapter to the commercialization of London parks, which contrast with US green spaces that probably have a longer history of private-sector involvement in green space governance, revenue generation, and management, and particularly botanic gardens. Smith concludes that there is a danger of over-exaggerating the "threat" posed by events but the increase in London particularly[4] has three important implications:

1. Events affect the symbolic, physical, and financial accessibility of green space.
2. Commercialization of parks makes them like the rest of the city, dominated by commercialism and construction, and jeopardizes the image of London being a green haven.
3. While temporary in nature, events provide precedents for further commercialization and additional events attracting more and more visitors, thus leading to the fact that they become destinations rather than amenities.

Smith (2019) concludes, and this surely has application to US and other gardens, that there is a need to prevent over-exploitation and over-commercialization of gardens, for greater input from stakeholders, particularly local stakeholders, and for greater transparency concerning the revenues and their use than is currently the case.

New Directions in Urban Garden Education; Fairchild Tropical Botanic Garden

In 2013 *Garden Tourism* showcased an innovative program put forward by Fairchild Tropical Botanic Garden in Coral Gables, Florida, called the Fairchild Challenge (Benfield, 2013). This program still exists and as of 2019, it could claim to be in 400 Miami-Dade County schools and reaching 120,000 K–12 students. However, Fairchild has gone beyond the challenge and initiated two projects that rank among some of the most daring and innovative in garden education and outreach.

The first initiative is entitled the Million Orchid Project and commenced in 2018.[5] More than a century ago, Florida was synonymous with North American orchids. Ninety-nine orchid species are native to Florida which constitute over half of the USA's orchid species. However, over the past century, native orchids were stripped from trees and sold as potted plants in the northern USA. Orchid populations dwindled to dangerously low levels[6] and with the rapid urbanization and development in Florida over the 20th century, many rare orchids only survived in wild remote places such as the Everglades. Successful growth of orchids is difficult and individualized, so reintroduction requires large numbers of growers and large areas with a significant number and density of trees with special propagation facilities. To that end Fairchild Tropical Botanic Garden is generating, through its propagation facilities, young native orchid plants to be placed in trees[7] in school landscapes, hospitals, and urban streetscapes where they can be interpreted for students, residents, and visitors. The goal is to have 1 million orchids growing in the greater Miami area by 2025. While this is an ambitious project in and of itself, Fairchild has extended its reach into the community with this program by adding the Million Orchid STEMLab. Recognizing that STEM (Science, Technology, Engineering, and Mathematics) education is required for the 21st student, in March 2019, in cooperation with Miami-Dade public schools and the University of Miami's architecture design/build program, Fairchild outfitted a yellow school bus into a state-of-the-art, mobile botanical micropropagation lab whereby middle-school students could propagate orchids as part of the Million Orchid Project. Students plant and transplant orchid seeds in the sterile conditions of the lab and once propagated bring the orchids into the school grounds where their growth and ecological interactions can be monitored. The program is fully integrated into the Million Orchid Project, yet in this case has middle-school buy-in to conservation and education that is planned to be extended to all of Florida and ultimately the USA. The interesting feature of this program from an educational perspective is that any grade level and any disadvantaged child on the disability spectrum can participate and the program has been shown to be powerful and empowering for all students who participate.

The second major initiative undertaken by Fairchild is a Growing Beyond Earth – Innovation Studio. In 2015, in the film *The Martian*, Matt Damon was stranded on Mars and was required to support himself while awaiting rescue. Growing plants in a Martian atmosphere for survival became Matt Damon's challenge. Maybe coincidentally starting in 2015, Fairchild collaborated with the National Aeronautics and Space Administration (NASA) to address the need for food production aboard spacecraft[8] with a project entitled "Growing Beyond Earth; the Innovation Studio". The studio located at Fairchild will be a makerspace where participants design and test food production facilities for use in space. The project began in classroom-based student and citizen science[9] and has expanded into the development of plant-growing hardware for use in long-distance space travel. The program went through four development stages over the past four years:

- Year 1 – technical feasibility, suggesting the varieties of plants that could be grown on the orbiting space center.
- Year 2 – place in 30 schools a growth chamber where five plant varieties were tested, based on astronaut needs, which were:
  ○ a large edible biomass;
  ○ small space requirement;
  ○ growth under low-energy lighting; and
  ○ must have a high vitamin content.

Students in the 30 schools were required to test these plants through a portal by collecting data (leaf size, growth rates, etc.).

- Year 3 – NASA came to the schools to standardize the data collection phase following tweets from the students in the previous year.
- Year 4 – NASA awarded Fairchild and the schools a four-year US$1.5 million grant to continue the program by testing over 130 varieties of plants.

The next stage is:

- Year 5 – test the robustness of these plants in space by flying them in space on a mission.[10]

The two programs above have also proven to be the driver behind other education programs and policies in the Greater Miami area. Both programs have stimulated interest in STEM, and particularly girls going into STEM disciplines, and thus validate the link between science and education. At a student level, it is clear these programs impact motivation, confidence, and achievement. This in turn has led to organizations like The Nature Conservancy approaching the garden and its school partners to examine things like the heat island in the city of Miami and the programs that might be started on a joint basis to address this issue.

Grow Wild; Royal Botanic Gardens, Kew, London

Out of its facilities in Richmond, London, Royal Botanic Gardens, Kew operates Grow Wild, a national outreach initiative which encourages greening community spaces and fosters community connections through sowing and growing wildflowers. Grow Wild, Kew inspires millions of people to grow as a group, get active, learn about and engage with nature, and give back through volunteering. Each year, Grow Wild funds around 50 community group projects across the UK that engage hundreds of people in the joy of plants and fungi in their local area. To date it has funded over 250 community-led projects with training and financial support of up to £4000 each. Through these projects the garden has reached thousands of people, often in poor or deprived areas, engaging them in the beauty and value of nature. As well as increasing the number of wildflowers grown across the country, Grow Wild aims to create wider benefits for society. These include bringing communities together in a joint endeavor, helping young people to connect with the natural world, promoting mental and physical well-being, and creating spaces for public interaction.

New Directions in Passive Garden Learning in Urban Botanic Gardens

Since their very inception gardens have been laboratories of learning. Moreover, unlike zoos and most museums, gardens are environments of immersive learning rather than environments of active learning or learning at a distance from the subjects. Finally, it has long been recognized that visitors to gardens come for a variety of motivations: relaxation and stress relief are often posited as the major motivators, but the very same garden must also cater to school groups, families, or in the case of UK, gardeners who just come for "a nice cup of tea". So, with one garden, but multiple motivators, the garden is charged with educating its visitors. The primary means over the years for imparting that learning is by means of signage. In one plant proximus sign, the species is named, identified as a relative of other similar or different plants, often labeled as from where it originates, and rarely a description of its useful or unique characteristics is given. Unfortunately, numerous studies have shown that visitors do not read labels – even less so retain the information imparted. This concept of interpretation has been a problem and challenge for many years in the museum and particularly the garden community. While a solution to this challenge has yet to emerge, a new interpretive Master Plan by the US Botanic Garden in Washington, DC uses a new and possibly more effective interpretation approach than previously attempted. It is a new direction in garden interpretation.

As a basis for the approach it takes the following as axiomatic:

- visitors have a limited attention span for the words on the sign, 60 words on a sign seems the optimum;
- some visitors will just scan or "graze" the signs;
- one idea per paragraph is mandatory;

- there should be a limit on the number of signs;
- the approach should be uniform for the whole building; and
- instead of building the exhibit around the collection, the interpretation should be retrofitted and the plants used to be the exhibit for the collection.

From these basic requirements, the traditional development process of asking the following questions may be addressed:

1. What are the main stories?
2. What are the main plants for that story?
3. What do the gardeners think of the layout/ selection, etc.?

Given the above, the US Botanic Garden decided on an interpretation scheme of "layering" in which the information would be broken down into layered "chunks". The layers consist of:

1. "Big ideas" such as evolution, global warming and plants, deforestation. These signs are heavy, semi-permanent, and prominent.
2. "Story labels" that group or highlight certain plants and must be viewed more carefully as to what they show or present. These signs must be changeable and movable (signs get wet, shade plants, and decay, hence they must be movable).
3. Scientific plant ID labels. Small, minimum words (30–40) showing plant family (in Latin, e.g. Laminae or mint family), scientific name, common name, and origin.

Within this approach the graphics have a singular look, there is a certain common look to the graphics, there is a pleasing yet unobtrusive color to the signs (in the case of the US National Botanic Garden, a nice blue/yellow pastel was used although reflectivity was an initial problem), and it forces the viewer to look at the individual plants. One outcome of this focus on plants is that individual "crowd pleasers" could be showcased. Thus in the primeval garden, dinosaurs can be shown (though NOT as an object of learning rather as a backdrop), carnivorous plants can be used, the titan arum or "stinking corpse flower" can be used to showcase tropical environments, and so on.

At the time of writing two test areas were in operation. In the tropical conservatory three large themes (big ideas) are presented:

1. What is this (tropical) habitat?
2. Rainforests in crisis.
3. Diversity in rainforests.

The second area is the desert or arid room where such concepts as convergent evolution are explored and plant adaptations to extreme heat and cold are explored on story labels alongside traditional plant ID labels. Of course, the next stage is the evaluation stage where the efficacy of the interpretation is assessed.

Sustainability in Urban Gardens

The term "sustainability" was first used in 1972 in the book, *A Blueprint for Survival* (Goldsmith, 1972), and in 1974 in the USA to promote a "no growth" economy.[11] The term was increasingly used in the 1970s and 1980s but gained considerable currency when used in the Brundtland report of 1987[12] advocating sustainable development. Today the term is widely used, and it would be rare to find a garden that does not specifically refer to, let alone claim to practice, sustainable development. *Sustainability* may be defined as: "The responsible use, maintenance and fortification of resources currently available such that future generations can derive the greatest possible benefit from what they inherit."[13]

Given that plants are ubiquitous on our planet and that much of the destruction and despoilment of the planet's resources has greatly affected the plant world, it is not surprising that botanic gardens are playing a lead role in the application of and education about the need for a sustainable future. Two examples may suffice: the first is drawn from Glenstone Museum and the second from Auckland Botanic Gardens, both catering to large adjoining urban populations.

Glenstone Museum, Potomac, Maryland, USA

Glenstone was chosen as an example of sustainability in gardens as it was the first visitor facility in tourism to aspire to 100% sustainable practices and as a result hired the first Chief Sustainability Officer for any garden entity. As the focus of the museum's environmental efforts, an environmental center was completed in 2019 and from which the surrounding 228 acres of woodland, meadows, and the museum infrastructure is managed. The major sustainability issues and resultant programs in the garden are sevenfold:[14]

1. Stream restoration. Glenstone, like many rural gardens, has within in it or borders significant watercourses. At Glenstone two tributaries of the Potomac River run through the property but before 2015 the streams were badly degraded. An action plan to address these concerns resulted in natural materials being used to stabilize the watercourse and the planning of appropriate aquatic vegetation resulted in lower sedimentation, stabilized banks, improved water quality, and a revived wildlife habitat for aquatic wildlife. The restoration also provided significant esthetic improvement to what had formally been a tire-filled, unsafe watercourse.
2. Reforestation. The 228 acres of Glenstone was formerly covered by Eastern Woodland species with a complete biome of arboreal and ground species harboring an active wildlife community. From 2013 until 2018 Glenstone planted over 7000 native trees, tens of thousands of shrubs, and millions of annual grasses and flowers. The result has been a site dedicated to native species and home to many returning wildlife and bird species.
3. Invasive species eradication. The literature on examples of invasive plants is extensive and far too numerous and complex to describe here. What might be said is that combating invasive species starts at the local level and Glenstone is part of that movement by initiating programs focusing on invasive plant eradication, research, and replacement, of which the installation of 30+ acres of native meadows, or 10% of the property, is a significant part.
4. Organic landscape care. At Glenstone organic chemicals for pesticide and herbicide applications are eschewed in favor of organic natural fertilizers derived from recycled ingredients in the waste stream. Furthermore, Glenstone's policies on watering restrict the casual application of water in favor of proper watering techniques.
5. Water management. The large area of most gardens and the imperative for water, both for the plants and for the esthetics, make water

management a major part of garden operation and thus sustainability. At its most simplistic, this is concerned with directing the water flow but also involves, in some gardens, water storage, flood plains, and bioretention areas. At Glenstone the need for water management has required six storage reservoirs – three underground and three above ground. In addition, Glenstone has provided for bioretention areas in the form of bogs, wetlands, and ponds that not only address flooding and drought but also provide habitats for insects and other fauna.

6. Composting. Glenstone composts all its organic waste, such that within 3 months any organic waste (including paper plates, napkins, and cups) is ready for use as organic supplements. Moreover, compost tea is made from composted organic material and is used as a fertilizer and a soil amendment throughout the grounds.

7. Recycling. Glenstone is committed to waste reduction whereby any waste will be recycled and in a 12-step program has as a goal the complete recycling of all materials that would otherwise be thrown away as trash. Montgomery County in which Glenstone lies had a goal of 70% reduction in waste by 2020. Glenstone challenged and met that goal in one year.

Auckland Botanic Gardens, Auckland, New Zealand

Auckland Botanic Gardens was chosen as an excellent and dedicated example of urban sustainability because it has been a leader in sustainable garden management for some time. The garden directs its efforts to three areas, as outlined on its website and elsewhere:

1. Sustainable horticultural practice "underpins everything we do. Our goal is to optimize the health of our plants without relying on fungicides or insecticides. We want to establish plants that are self-sustaining, and require minimal intervention. To achieve our goals, we use plant trials to identify plants that perform well and remain healthy in Auckland's mild climate, without pesticides. In conjunction with wise plant selection we also focus on soil health to help bolster plant resistance to disease. We increase soil health by covering soils with mulch and adding compost when planting".[15]

2. "A particular focus is given to the role plants play in environmental enhancement, for example, a Water Sensitive Design (WSD) to protect and enhance our waterways is in place whereby stormwater treatment from not only the garden but outside is channeled through various devices to provide water for the garden and thus decrease water needs. For visitors, a 'Sustainable water trail' has been signed to explain water management in the garden".[16]

3. "The third practice Energy Conservation, by emulating the plants they grow by harnessing the power of the sun. We've started out with a few solar and PV panels to run our Education Centre".[16]

This emphasis on sustainability has been recognized and extended by the garden hosting Auckland Council's Education for Sustainability program, whereby the garden runs Learning through Experience programs as part of the curriculum for 0–13 student grades. Finally, Auckland Botanic Gardens was leader in the conservation goal of emphasizing regional and nationally threatened native plants and through their cultivation and production, disseminate the plants either physically to local gardens and gardeners or often in the restoration of wild populations, through using their education programs and seed banks to ensure the plants' survival.

Urban Gardens and Changing Ethnicity

The Brookings Institute in 2011 wrote: "The substantial racial and ethnic changes in the population of both cities and suburbs in metropolitan America, challenge leaders at all levels to understand and keep pace with the continuing social, economic and political transformation of these places" (Frey, 2011). The study, based on the 2010 census, suggests five major demographic changes taking place in the USA. They are:

- well over half of America's cities are now majority non-white;
- Hispanics now outnumber blacks and represent the largest minority group in major American cities;
- minorities represent 35% of suburban residents, similar to their share in the overall US population;

- more than half of all minority groups in large metro areas, including blacks, now reside in suburbs; and
- fast-growing ex-urban areas remain mostly white and depend overwhelmingly on whites for growth in the 2000s.

These dramatic shifts have major repercussions for gardens, as most gardens, owing to urban growth, are now within city limits or suburban areas and new gardens like Houston, Delaware, and Pittsburgh are required to adjust to new ethnic audiences in order to remain relevant and grow. Several gardens had to adjust to this new reality at an early stage.

Case Study 9.1: Queens Botanical Garden, Flushing, New York

Queens Botanical Garden owes its origin to a 5-acre site called Gardens on Parade as part of the 1939 New York World's Fair. The garden was saved from destruction in 1946 as the garden of the Queens Botanical Garden Society and in 1961, it was moved to its present 39-acre site on Main Street, Flushing, as part of the neighborhood redevelopment for the 1964/65 World's Fair in Queens. In its earliest years, the Flushing neighborhood was almost exclusively white, including a sizable Jewish community, but in the 1990s Chinese, Korean,[17] and Latin American immigrants moved into the area, and even more recently immigrants from such other countries as Burma, India, and Nepal found the borough a desirable place to live. Today the ethnic mix is one of the most diverse in the USA and the garden is in the middle of that ethnic neighborhood. While the garden has always functioned as a desirable place for all ethnic communities, the need for greater earned income,

particularly in the period of diminished grants from New York and its granting agencies, required the garden to not only start charging admission but also provide programs that catered to the ethnic community surrounding it. The programs mostly take the form of specific events targeted at specific groups. At times the garden has had very large events, with the grounds often rented by the specific cultural communities and corporations. It is known that people love to celebrate outside – to enjoy nature, the sun, their food, music, each other. There have been rentals for Buddha's birthday, Punjabi Mela, Feria de las Flores with the Colombian community, the Pakistani community, and on September 7–8, 2019 the garden hosted a rental/public program for the Korean Moon Festival, which is very important to Koreans!

The garden also reaches out to other community partners. Uber rented it a couple of years ago to show appreciation for its local customer base.[18] The local hospital sponsored its winter holiday party at the garden and some other doctors have rented the garden for a department outing. H-Mart, the well-known Korean grocery chain, sponsored a picnic where kids got backpacks full of food and were invited to do a drawing and enter a contest. The art of the winners was used in the company's marketing materials. In 2018 the garden launched Birthday Parties which have been very successful, with 12 in the first year, and the goal is to double that number for 2020. All the children get butterfly wings, help make food with herbs from the garden, and tour the garden. Over the past 12 years there have been three Leap Year Birthday Parties (2020, 2016, 2012) at the garden[19] and on February 16, 2016, the garden featured a Quinceañera, complete with a Mariachi band and the traditional "changing of the shoes" when young ladies turn in their child shoes and are given high heels.[20]

Notes

[1] Including many movies in the *James Bond* series, and the remake of *Mary Poppins*.
[2] For example, coffee houses have opened in some parks (Perkins *et al.*, 2009).
[3] In Battersea Park, London, Go-Ape attractions have been built.
[4] Smith (2019) notes that Battersea Park has gone from staging 100 events in 1991 to over 600 in 2015.
[5] It was modeled on a program initiated by Singapore Botanic Gardens.
[6] For example, it is estimated there are only 25 cowhorn orchids (*Cyrtopodium punctatum*) left in the wild.
[7] Surprisingly, Fairchild found the orchids actually grow better in urban environments than in other Florida locations.

[8] While the program must be considered innovative and out of this world (pun intended!), the model originally came from Polynesian explorers from whom one can trace long-distance Pacific exploration by means of "paths by plants along the way".

[9] Today it is in 100 classrooms.

[10] As an answer to the question everyone asks, it seems two plants hold the most potential for long-distance space travel use: extra dwarf bok choy and dragon lettuce.

[11] In 1974 a recession brought on by an oil shortage made the application of the no growth economy a real concern in the USA. For a summary of this period of no growth, the reader is referred to https://www.emerald.com/insight/content/doi/10.1108/eb053717/full/pdf?title=the-no-growth-economy (accessed July 29, 2020).

[12] Entitled *Our Common Future* (World Commission on Environment and Development, 1987).

[13] This definition is drawn from the *Glenstone Field Guide* available to visitors at the new Glenstone museum, 45 min north of Washington, DC, and reflects Glenstone's outlook on management and development. To this definition, and in practice, Glenstone believes that remediation and repair should also be included, given the status of much of our planet.

[14] Available at: https://www.glenstone.org/wp-content/uploads/prod/2019/10/EC-Digital-Presentation.pdf (accessed August 4, 2020).

[15] Available at www.aucklandbotanicgardens.co.nz (accessed May 26, 2016).

[16] Signage in the garden (May 26, 2020).

[17] Numbering over 60,000 in 1990.

[18] Uber as a ride-sharing operation is most in demand during rainy days; the fact that it poured with rain all that day was coincidental, but a good time was had by all.

[19] In 2020, February 29 was a Saturday, so large crowds were expected. The coronavirus greatly diminished participation that day.

[20] Garden staff naturally received green garden clogs!

Flyer circulated to residents and visitors concerning the possible construction of two 39- story towers next to Brooklyn Botanic Garden with the potential to block significant sunlight. Author's own photo.

Making Brooklyn Bloom is a program run by the Brooklyn Botanic Garden to beautify the borough of Brooklyn. A seminar on the program is provided at the start of every year at the garden. Author's own photo.

A modified school bus is used by Fairchild Botanic Garden to go to Miami schools to deliver the million orchids program. Author's own photo.

The Green Flag Award is a prestigious award given to gardens of outstanding quality. Westminster City Council in the heart of London had 31 gardens which were Green Flag winners in 2019. Author's own photo.

Victoria Embankment, Westminster, London is the location for workers, residents and tourists to enjoy floral splendor often in association with entertainment and events. Author's own photo.

Finsbury Park, London. A public park fenced off for an event said to be highly indicative of Neoliberalization affecting parks and gardens. Photo by Dr Andrew Smith, University of Westminster.

Crowds waiting to enter the Wireless Festival at Finsbury Park. Photo by Dr Andrew Smith, University of Westminster.

Signage, its readability and content is a major consideration when fulfilling and imparting its educational mission. This sign is at the US National Botanic Garden, Washington, DC. Author's own photo.

Blithewood Garden on the Bard College campus in Annandale-on-Hudson, New York. The Garden Conservancy is partnering with Bard to rehabilitate and renovate the decaying architectural structures of this gilded-age garden, part of which will be selected tree removal to create the former view across the Hudson River. Photo courtesy of the Garden Conservancy.

10

Gardens and Historic Homes; New Directions in Historic Garden Tourism

_____

DOI: 10.1079/9781789241761.0010

Gardens are invariably associated with a residence and in many cases of garden tourism, that residence is a historic structure. The link between domestic tourism, gardening, and visiting historic sites has been documented in the UK by the National Garden Scheme (Buck, 2016) in which not only the growth in gardening as an activity was shown (Fig. 10.1), but also the strong link between days out and visiting historic sites by gardeners (see Fig. 2.2). There has been an increasing interest in history and historic sites in the USA since 1980 as shown by the rise in visits to all historic sites, where in 2018 there were some 113 million visits (Fig. 10.2). Statistica indicates that in 2014 "visiting a historical site" was the third most popular activity while on vacation after "shopping" and "swimming/water sports".[1]

While participation in "visiting historic houses" is chosen by approximately 14% of American tourists when on vacation, research has shown that when the historic structure has a garden associated with it, the majority of visitors stay in the garden and do not tour the house. The converse, that where there is a (historic?) house associated with the famous garden, few people tour the house in preference to touring the garden. Thus, at Stourhead in Wiltshire, England, approximately 10% of visitors tour the house with the majority touring the landscape garden. At Chatsworth, the more famous stately home of the Duke and Duchess of Devonshire, only 25% pay to go into the house with the other 75%, out of a total visitation of 850,000, staying in the garden and grounds. At the more than 300 National Trust properties that charge a fee for entry, gardens drive visitation more than

houses and yet when visitors are asked why they visit the National Trust property, the first three reasons are not related directly to the garden or historic site; the top five responses are:[2]

1. For "beauty and tranquility".
2. "For a walk".
3. Access to a tearoom.
4. An interest in National Trust gardens (50%).
5. Of whom, 50% say they are interested in the garden being visited while the other 50% say they came to the property to relate to their own garden, presumably for design and plant advice/ideas.

As was noted in Chapter 2, owing to a decline in satisfaction at National Trust properties, the Trust has embarked on a series of new initiatives, mostly facility- and infrastructure-based, to address this problem. While the improvements would be too numerous to note herein, National Trust surveys show that visitation increases significantly, sometimes over 100% in volume, once new or better infrastructure is provided. As a result, there has been a marked increase in interest in renovating historic properties of all kinds in the past 5 years. The renovation has essentially involved three new directions.

• The first new direction that has become increasingly apparent over the past 5 years has been the renovation, conservation, and preservation of the historic portion of the tourist destination, whether it be a garden or a structure.

• The second new direction over the past 5 years has been the restoration of historic gardens and properties that have fallen into disrepair

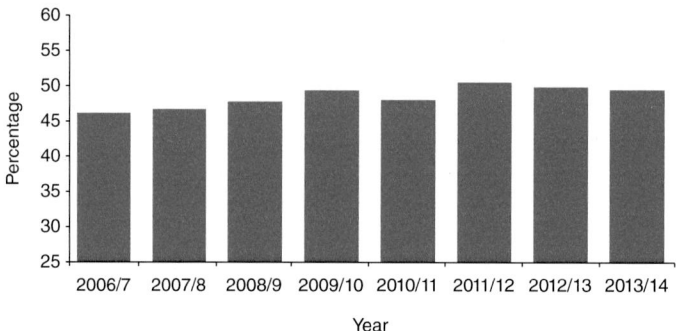

Fig. 10.1. Percentage of adults in England reporting gardening as a free-time activity, 2006/07 to 2013/14. Adapted from Buck (2016).

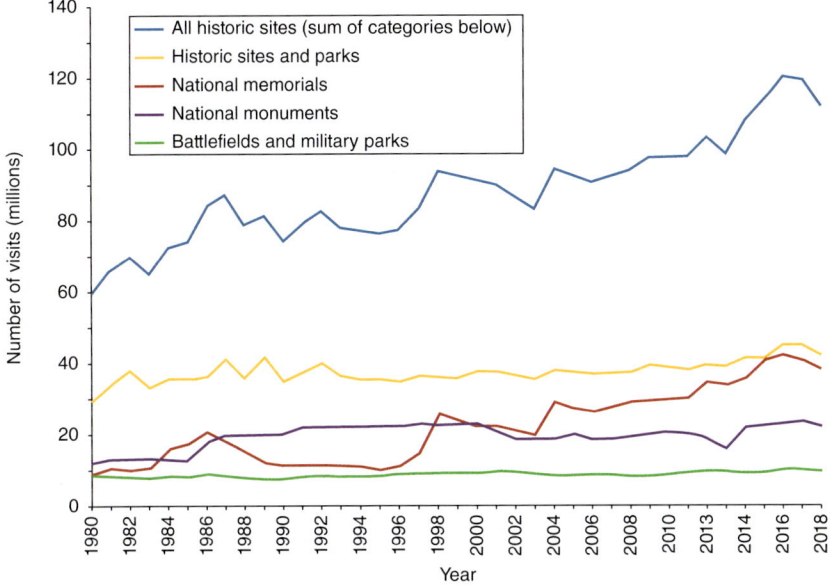

Fig. 10.2. Visits to historic sites in the USA, 1980–2018. Humanities Indicators, 2019; adapted from American Academy of Arts & Sciences (2019).

or that have since been seen as valuable and in need of restoration.

- The third area in which historic (gardens) tourism has been reevaluated is in what has been called the "historic buildings and landscape collection movement" in which historic buildings and their contents no longer become the focus, but a movement to where the (changing) landscape in which the collection is situated becomes integral to the understanding and interpretation of the historic site. At the heart of this concept is the recognition of *place* as a defining principle of historic structure and landscape interpretation.

Integral to the changing and new directions in historic properties is the different and distinctive strategies that are applied to historic properties. In this chapter, examples will be presented of the four strategies of:

- preservation;
- reconstruction;
- restoration;
- renovation;

and the fifth new direction of:

- landscape integration.

Prior to examining these strategies, their definitions and thus the differences between them are first presented, as follows (adapted from The Burra Charter, 2013):

- *Conservation* means all the processes of looking after a place to retain its cultural significance.
- *Maintenance* means the continuous protective care of the fabric and setting of a place, and is to be distinguished from repair. *Repair* involves restoration or reconstruction.
- *Preservation* means maintaining the fabric of a place in its existing state and retarding deterioration.
- *Restoration* means returning the existing fabric of a place to a known earlier state by removing accretions or by reassembling existing components without the introduction of new material.
- *Reconstruction* means returning a place to a known earlier state and is distinguished from restoration by the introduction of new material into the fabric.
- *Adaptation* means modifying a place to suit the existing use or a proposed use.
- *Use* means the functions of a place, as well as the activities and practices that may occur at the place.

- *Compatible use* means a use which respects the cultural significance of a place. Such a use involves no, or minimal, impact on cultural significance.

Different Strategies Applied to Historic Properties

Garden tourism and preservation

The State of Virginia is the location of the first permanent white settlement in America, in Jamestown in 1607. The original colony did not last but the colonists from England expanded out from that first settlement and by the 20th century, Virginia was home to large plantations with large antebellum homes and invariably with a splendid garden. Places like Monticello (Thomas Jefferson's home), Mount Vernon (George Washington's home), Bacon's Castle (the oldest brick building in the USA), Lewis Ginter Botanical Garden in Richmond (home of a wealthy 19th century industrialist), and the State Arboretum in Winchester presented perhaps the largest collection of historic gardens in America. The 20th century saw a decline in the fortunes of many of these homes and in 1929 the Garden Club of Virginia (GCV)[3] embarked on a program to preserve, and in some cases restore, the state's historic public and private gardens. One of the means used is the holding of the Historic Gardens Week in Virginia. Started in 1929,[4] and now the nation's oldest house and garden tour event, Virginia's Historic Garden Week is the opening of 250 gardens (and often the associated homes) over an 8-day period every year in late April/early May. It caters to over 25,000 visitors (it is a ticketed event) with suggested tour itineraries and stops courtesy of a 213-page guidebook.[5] Every year the homes and gardens showcased change, making each Historic Garden Week a unique experience.

Most importantly, the money raised is used almost exclusively for preservation and restoration of the state's historic gardens. In 2014 Historic Garden Week proceeds funded the restoration and preservation of more than 40 of Virginia's historic public gardens and landscapes, a research fellowship program that documents significant gardens, and a 5-year project with Virginia State Parks, in celebration of the GCV's Centennial in 2020. In the same year, 2014, an economic impact assessment was made of the historic garden tours' monetary contribution and impact on the state's economy. While complete data were not available as to the very first tours in 1929, the GCV was able to provide figures estimating the cumulative impact of Historic Garden Week over the last 45 years to be US$425 million. In 2014 alone, nearly 30,000 visitors came to 234 private houses, gardens, and historical sites that comprised its 31 tours around the state. Data were collected for visitor spending (60% of tourists have an income level over US$100,000) on food and hotels (25% of visitors stay at least once overnight spending an average of US$1207 on their garden week trip), as well as the costs of preparing homes and gardens for tour day (which often begins several years in advance)[6] and the financial benefit that the GCV's garden restoration projects reap from the annual week-long event. Total economic impact that year was estimated at US$11 million (Chmura Economics & Analytics, 2016). One benefit of this form of garden tourism is that it highlights and provides economic benefit to both rural and urban communities across the Commonwealth of Virginia.

Garden tourism and reconstruction

Perhaps the most thorough garden reconstruction and one of the oldest is the restoration in 2009 of the Elizabethan Garden at Kenilworth Castle in Warwickshire, England. Dating from the 12th century, the castle was an important stronghold and then became the seat of an earldom and one of the most important sites in Elizabethan England when Robert Dudley, Earl of Leicester, acquired it in 1553. It was the site of a 19-day stay by the court of Elizabeth I and was lavishly remodeled, renovated, and added to, to impress the queen in the hope that she would marry Dudley. Among the additions was an Elizabethan garden in the Italianate style with aviaries, knot gardens, wooden obelisks, a central marble fountain, and sculptures. In 1651 with the coming of the Civil War, Kenilworth Castle was slighted so that it could not present a threat to whomever won the war and over the next 400 years the garden was abandoned, and the layout lost. In 2005, English Heritage began a program of restoration that cost over £3 million and that was finished in 2009. The reconstruction was based on historical archeology and the Langham letter dating from 1580, that describes the nature of the garden in some detail, and the garden is now a major feature in English Heritage's avowed goal "to bring the story of England to life" – in this case to 1,100,000 of the 10 million visitors to English Heritage properties in 2017.

Garden tourism and restoration

For almost 30 years the name Alcatraz was synonymous with the most secure and famous penitentiary in the world. Located on an island in San Francisco Bay, California, it held the worst of the worst criminal offenders in the USA – Al Capone, Alvin "Creepy" Karpus, Robert Stroud (the "Birdman of Alcatraz"), and George Kelly Barnes ("Machine Gun Kelly") – hardened criminals who would seem totally divorced from gardens and gardening. But soil had been brought to the barren island as early as 1870 and gardens cultivated firstly by personnel from the military prison were expanded by staff and families when it became a federal penitentiary in 1934. The gardens expanded significantly in the 1930s when horticulturalists were consulted as to what to plant on a water-deficient island and prisoners were permitted and encouraged to further terrace the slopes and expand the garden area. A greenhouse was even built to provide plants and seedlings to the extensive gardens. The prison closed in 1963 and the gardens fell into such disrepair and abandonment that by 1990 the former garden areas and terraces were overrun with wild berries, ivy, and weeds.

In 1990 the Golden Gate National Parks Conservancy launched a restoration drive, but it was the Garden Conservancy that accelerated the drive for full garden restoration. A US$250,000 grant in 2003 funded a cultural landscape inventory and in 2003 volunteers began the work of restoring the gardens including the warden's gardens, the prisoner's garden, the terraced slopes under the cellblocks, and even the rebuilding of the glasshouse. In 2006 a dedicated garden brochure was added for visitors and garden tours with trained docents commenced in 2007. Information panels were introduced in 2009 and by 2010 the gardens had become a major media attraction alongside dedicated souvenir commemorative books and a Facebook page. Today, the gardens are an integral part of the story of Alcatraz, namely one of incarceration, inspiration, survival, and human determination. In a wider context the gardens are a major part of the visitor experience to San Francisco and the island of Alcatraz. In 2018 over 26 million visitors came to San Francisco and 25.7% visited Alcatraz.[7] This made Alcatraz the fifth most popular tourist destination in the city after Pier 39 at Fisherman's Wharf followed by the Golden Gate Bridge, Golden Gate Park, and Lombard Street – or put another way, the island of Alcatraz is the most popular paid destination in the city.[8] There are no data on visitors to the garden specifically, but every guided tour includes the now-restored garden.

Garden tourism and landscape

In the history of garden design, landscape itself and design elements within the landscape represent a highly significant period in English garden design history. Lancelot "Capability" Brown, William Kent, and Humphrey Repton are the most well-known English landscape designers of the 18th century but their designs fell out of favor in the subsequent century with the rise of the romantic period and the preference for and promotion of wild untamed nature. In the development of historic houses, *period* became the guiding principle and houses were set in the context of the period in which they presumably reached their apogee. The best example most often cited is 18th century Colonial Williamsburg, Virginia with restored and refurbished colonial buildings. More recently the concept of stagnant frozen-in-time homes has been questioned (Zar, 2016) in favor of a more dynamic, changing interpretation and with a recognition that the landscape in which the house was placed and evolved is more realistic and authentic than the earlier period presentation and interpretation.

Two examples may suffice. In New Canaan, Connecticut, the architect Phillip Johnson built a glass house in 1948 that today has become a very popular and, for some, iconic structure. It is open to the public and receives over 13,000 visitors per year, many being modern art aficionados and architects. However, what is little known is that Johnson saw the house as one element in a landscape and in the manner of the English 18th century landscape designers, planned and cut vistas, created follies, and planted trees within an ordered and artistic landscape.[9]

The second example comes from Lyndhurst Mansion in Tarrytown, New York, in the Hudson Valley. Lyndhurst was started in 1838 but it was the second owner, George Merritt, who substantially expanded the grounds by hiring the landscape designer Mangold who planted lawns and trees with distinct views and curving driveways for "surprise' vistas and built a conservatory, some of whose plants started the NYBG's collection in 1900. Following Merritt, the estate passed into the hands of the financier Jay Gould and his descendants; on the death of Gould's last daughter in 1961 the estate passed to the National Trust for Historic Preservation. In the years of the Gould family ownership the grounds became overgrown, structures rotted and disintegrated, and plantings were abandoned. In 2012 a Master Plan was adopted to reestablish the landscape[10] and since then rookeries have been exposed, paths reestablished, trees planted, and rebuilding of the seating at the scenic overlook of the Hudson River. The next step is to refurbish the swimming pool as an event space and restore the rose and perennial gardens. Restoration of the greenhouse is a longer-term goal owing to cost, utility, and repurposing if necessary or advisable.

Restoring and reimaging a garden for garden tourism

In 1977 the Garden Museum on the south bank of the Thames and adjoining Lambeth Palace was opened. It is located in the Church of St. Mary-at-Lambeth, a church dating from the 10th century and that significantly contained the tomb of John Tradescant Jr. and his father, John Tradescant Sr., who was considered the father of British gardening. As gardener to Queen Henrietta Maria, Tradescant Sr. traveled as far as the Arctic and the Sahara Desert to collect and classify plants; and his son journeyed to America to collect plants – the first known botanist to catalog American plants. In total the two traveled to three of the four known continents at the time. Both are buried at St. Mary-at-Lambeth, with John Jr. buried in an elaborate tomb in the garden of the church, hence the importance of the site to garden history.[11] Under the threat of demolition, the site was saved as a garden and the Garden Museum opened in 1977 to showcase the history of gardens. In 1980 an Elizabethan knot garden was created as a quiet reflective space in the center of a bustling London. In 1990 plans were made to expand the museum and showcase the artifacts that had been collected since the museum's opening. The museum was attracting 40,000–50,000 visitors in 2012/13 and it had the goal, with the refurbishment, to attract 75,000 visitors. In 2017 the museum reopened and in 2019, its first full year of opening, claimed a 400% increase in visitors.[12]

In 2015 the museum closed for two years and was renovated with the following features:

- the medieval tower was stabilized and reopened for visitation;
- the interior was gutted (all pews and church infrastructure removed to permit display space);
- the upper mezzanine was converted to display space;
- the knot garden was removed and replaced by a modernist cloister garden at the center of which is the Tradescant tomb;
- an upscale café facing the street and cloister was installed;
- a second, new garden was installed in the front of the museum/church that is meant as a resting, waiting, and entry space into the museum from the traffic just meters from the church;
- historic gardening tools and a 17th century water pot are among the 100 or so artifacts on display;
- the Tradescant ledger stone, a simple engraved slab inlaid in a church under which the deceased is buried; and
- a book of "rarities", the record of the plants the Tradescants collected and cataloged, is the major item now on display.

In 2017, when the museum had been closed for two years, during the lifting of some of the ledgers in the church, a 15th century crypt containing over 30 tombs was discovered under the church floor. While currently inaccessible to the museum visitors, this represents a major addition to the history and attractions of this unique garden museum.

Case Study 10.1: Bellamy-Ferriday House & Garden, Bethlehem, Connecticut, USA

Connecticut Landmarks was founded in 1936 and has ten historic structures and often gardens that span four centuries of Connecticut history. One of three houses, the Bellamy-Ferriday House & Garden, in Bethlehem, Connecticut, was bequeathed to the society in 1990 and for many years hosted about 3000–5000 visitors to the house and the garden which was designed and planted by Mrs. Ferriday and her daughter Caroline following their acquisition of the property in 1912. The garden is particularly notable for its formal parterre garden with roses, peonies, and most recently and most notably, its lilacs. During World War II Caroline Ferriday worked at the French Consulate in New York City and as a member of the French Resistance movement, work for which she received the Cross of Lorraine and the French Legion of Honor. She was particularly interested in the story and fate of the "lapins" or "rabbits"; women interned in the Ravensbrück Nazi concentration camp and who were subjected to medical "testing". She began a crusade in 1957 to bring the 35 surviving women[13] to the USA to have restorative medical attention. It was through a historical fictional novel entitled *Lilac Girls*[14] that the story of her efforts was told by Martha Hall Kelly. The novel

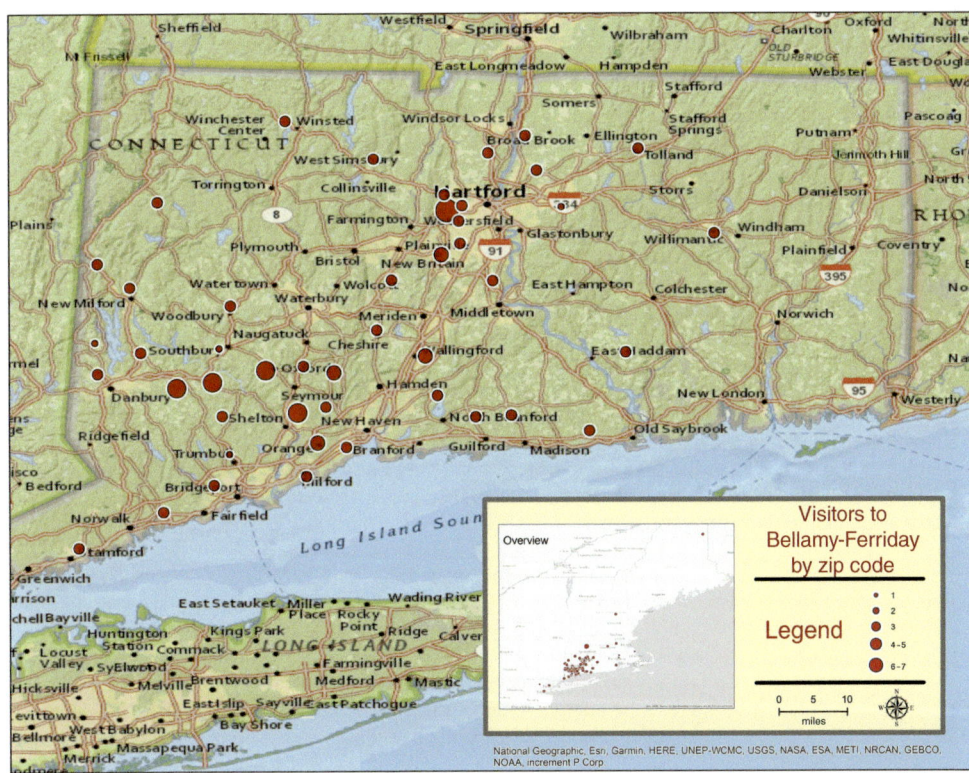

Fig. 10.3. Choropleth map of visitors to Bellamy-Ferriday House & Garden, Bethlehem, Connecticut, May–July 2019. Map courtesy of P. Meng and T. Wendeborn, CCSU Department of Geography.

was a success, going immediately on the *New York Times* best-seller list in April 2016 and has remained there up to the time of writing.[15] The fact that Ms. Ferriday lived in Bethlehem and the lilacs she planted were such a focus of the book created a surge in attendance at the house and garden. Bellamy-Ferriday recorded 4739 visitors during 2019, an increase of over 20% from 2018 which saw 3728. The following results are part of a visitor origin study[16] applied to visitors in the summer of 2019.[17] The survey showed 66% of the visitors had heard of the house and garden through the book and a remarkable 80% of visitors knew of the significance of the house from the book. Sixty-six percent of the visitors specifically said they had come because of the book *Lilac Girls* and 66% had read the book. Of the respondents, 39% were over the age of 60 years and 64% over the age of 50 years, 86% were

female, and 54% belonged to a book club but only 14% to a garden club. However, 98% of the respondents gave the garden a rating of 5 out of 5. When asked what influenced their decision to come to Bellamy-Ferriday House, 66% had come because of the book. Bellamy-Ferriday also conducted a survey of *where* the visitors to the house came from in the months of May, June, and July 2019. Fig. 10.3 indicates the zip codes of the visitors; it shows that a large number of visitors came from Connecticut, west of the Connecticut River, but a notable number from outside of Connecticut.

Not coincidentally, Ms. Kelly's next book was a prequel to the *Lilac Girls* entitled *Lost Roses*,[18] which uses Caroline's mother, Eliza, as the main figure and her roses as the symbol of the tragedy and tribulations of three women during World War I.

Notes

[1] See https://www.statista.com/statistics/298705/most-popular-summer-vacation-activities-for-travelers-us/ (accessed August 17, 2020).
[2] The response that focuses specifically on visiting National Trust "properties" – presumably fixed-roof facilities – ranks seventh.
[3] The GCV is one of the oldest in the USA. The first was in Athens Georgia in 1891. Since 1920 the GCV has grown from eight founding clubs to 47 clubs with over 3400 members. It is the coordinated efforts of these volunteers, along with the generosity of over 200 private homeowners across the Commonwealth of Virginia, who make Historic Garden Week possible.
[4] The first tours were organized to support restoration work at Kenmore, George Washington's sister's house, in Fredericksburg.
[5] Seventy-five thousand guidebooks are printed and are read by over 1.6 million readers.
[6] Fifteen tons of mulch is used by homeowners to prepare their gardens for this statewide event.
[7] San Francisco CVB tracks visitor volume into the city with the help of multiple data providers, plus intercept surveys to guage what visitors are doing while they are there. CVB reports 6.7 million visitors to Alcatraz per year (this would not include locals).
[8] There is some discrepancy in the numbers of visitors to Alcatraz. The tourism bureau has a figure of 6.7 million but information from the company that operates the ferry transportation to Alcatraz gives a number of 1.8 million because, based on National Parks Service restrictions, the ferry service is limited to about 2.0 million visitors to the island per year. Last year the number was 1.8 million people and was sold out about 95% of the time.
[9] It is said Johnson would sit on his veranda with binoculars and a walkie-talkie dictating which branches of trees needed to be removed or trimmed and what species of tree or shrub should be planted and where.
[10] Along with stabilization of the house including a new roof and conversion of the carriage house into a visitor center.
[11] Perhaps as significant, the garden also contains the tomb of Captain William Bligh – of "Mutiny on the *Bounty*" fame but also a key figure in the attempt to bring breadfruit to the Americas in 1845. Today breadfruit is important in many Pacific cultures and Hawaii has a Breadfruit Institute dedicated to the use and propagation of the fruit.
[12] Although the base number or year for the increase was not given; see https://www.nordiccopper.com/copper-stories/garden-museum/ (accessed August 14, 2020).
[13] Of the original 73.

[14] The author notes the title, *Lilac Girls*, comes from when Caroline explains to Kasia, when they are weeding in the garden, that lilacs only bloom after a harsh winter, just like the girls – they are blooming after their own harsh times.

[15] In July 2017 the book was still at #4 on the list.

[16] To avoid confusion, the questions asked, sequentially, were:
  1. How did you hear of the house?
  2. Why did you come today?
  3. Were you aware of the significance of the house
  4. Have you read the book?

[17] The survey was unpublished.

[18] Of which visitors, 31% had already read *Lost Roses*.

Line-ups characterize the houses in historic Richmond, Virginia during Historic Garden week throughout Virginia. Author's own photo.

Visitors stroll the gardens of an antebellum mansion in Tidewater, Virginia. Author's own photo.

The House of Seven Gables, made famous by American author Nathaniel Hawthorne's novel *The House of the Seven Gables*, is a 1668 colonial mansion in Salem, Massachusetts. The garden is toured as much as the house. Author's own photo.

Dedication ceremony of new greenhouse set up by volunteers at the Gardens of Alcatraz, San Francisco, California. Photo by Elizabeth Byers.

The High Line in New York City was formerly a disused rail line serving the west side of Manhattan. Last used in 1985, the derelict line was saved from redevelopment in 1999 as an elevated walkway and planted as an urban garden. Today more visitors walk the High Line than visit the Empire State Building. Photo courtesy of Jim Charlier, Charlier Design, and used with permission.

11

Future Directions

_____

So, now we have looked at new directions in garden tourism over the past 7 years or so, it is time to assess where these new directions have led us and where they will go over the next years. It is useful to start or reiterate where we left off in 2013, namely the data suggest that *garden tourism is today, and will continue to be, the dominant mode of outdoor leisure activity in the world.*

The corollary to this is that tourist economies that recognize and react to this statement may make it the most important emerging tourism activity in their country, because gardens appeal to all demographics, all the world's population, and all cultures in the countries in which they live.[1] The following four paragraphs present the most compelling and relevant data that permit this conclusion to be reached:

1. In the world today there are probably over 5000 public and private gardens, visited by over 1 billion people. This means that one in seven of the world's population visits a garden yearly. At a growth rate of over 7% over the past few years (usually weather dependent), it is the fastest-growing segment of today's tourism industry (in contrast, large ship cruising is growing at 4% and theme park visitation is also growing at around 4% per year). Most countries have many gardens both in urban and rural areas and a large variety in design, type, and type of attraction in the garden. Garden visiting is enjoyed by people of all ages, beginning in their early twenties, and never diminishing. It is enjoyed by all demographic types, all genders, all ethnic groups, all household compositions, and all income levels.

2. There is a wealth of evidence that gardening specifically, but also garden visiting, is good for a person's health and welfare. Gardens provide stress reduction, family bonding, and personal fulfillment, and furthermore there is evidence that gardens provide medical benefits in the form of increased happiness, pleasurable sensory stimulation, and personal peace and satisfaction.

3. Garden visiting is strongly associated with the movement toward sustainable living arising from climate change, food security issues, and the desire for environmental sustainability. Most gardens as part of their mission have an educational component that is targeted toward the understanding of the plant world for our sustainable future, if not survival; without plants there is no food, no alcohol, and no clothes, and plants provide the means for affecting climate change through trapping carbon and hence reducing global warming.

4. Finally, gardens have an important role in the economic viability and sustainability of communities, regions, and nations. The number, size, and density of gardens and the fact that they are found in both rural and urban areas make them economic generators for all sectors of the economy. For example, the world's pre-eminent botanic garden – the Royal Botanic Gardens, Kew – is found in the suburbs of London. The Eden Project, one of the most iconic gardens in the world, is in a disused quarry in the far southwest of England. There are botanic gardens in the heart of London in Chelsea, in Oxford, Cambridge, Bristol, Glasgow, Dundee, Belfast, and the Isle of Wight. Added to that, National Trust gardens are found in every region of the UK. There is no other outdoor leisure activity in the UK that can claim this distribution or coverage. Similarly, in the USA, major cities like Houston, Texas; Louisville, Kentucky; and Pittsburgh, Pennsylvania are all building new botanic gardens. Economic impact studies are available for many of these gardens and the numbers are significant, impressive, and (I believe) often understated.

Given the above evidence and importance, it is a mystery to the writer as to why gardens have received so little attention by academics, policy makers, and tourism marketing and development officials over the years. It is not important herein to understand why garden tourism has been, and still is, so often neglected; what is more important is to recognize the importance of garden tourism and work on it.

The introduction above can only suggest *why* garden tourism will be so important in the coming years. Following are ten areas that have been identified by various garden experts as new challenges in the directions noted above:

1. The first observation is that gardens, and particularly botanic gardens, have adjusted to the changing paradigm, outlined in the first edition of *Garden Tourism*, of being institutions specializing in botany and horticulture moving into the realm of the visitor and tourist business. It was probably around the turn of the 21st century that gardens were in the process of, or being forced to, change from *ex situ* museums for plants with little consideration for the visitor into a sector of the tourism industry that was interesting, relevant, and had its own niche in the tourism industry. If we reflect back on the thesis by Peter Drucker that was introduced at the start of this book, we might conclude that in a sense gardens embraced at this time a broader vision by examining what business they were in and came up with the answers for product development of botanical science, visitor services, and education. This of course answered Drucker's question of what business gardens were in, but it leaves unanswered the question of how this was to be marketed. In 2018, Kim Ellis,[2] as the Director of the Royal Botanic Garden, Sydney, indicated that all the world's gardens should answer the question: What are we delivering? The answer to that question is not as simple as a statement that the mission of the garden is the furtherance of science and plants. While a garden might create engagement on that story, the delivery message must also be communicated to legislators (and enshrined in legislation), policy workers, voters, funders, and visitors. At the strategic level this is concerned with communicating the *value* of the botanic environment. It is believed that this message, in the current political climate, is one of the few that has "traction", or as stated in Chapter 6, the brand may very well be:

* no alcohol
* no clothes
* no food

Brand: No plants/no life

And how do gardens manage this communication: What are gardens communicating to the visitors about what gardens are? If there are few to no "rules", do people think it is a free-for-all? If gardens have rules of conduct in these horticultural spaces, does that elicit a different response from the visitors? Do they take the space more seriously, or do they just choose not to go?

2. In the process, the importance of funding and the sources of that funding became important. In those past 20 years there has been a decrease in government funding and hence gardens were forced into looking at grants, earned income, and even the community in which they were located. The more proactive gardens realized that this change of focus required a broader set of skills from management and workers.

It was therefore not unusual to find garden managers needing training in property development, financial management, and asset management. Ellis suggested one new area gardens could increase their relevance was in the field of creating financial sustainability.

In the Royal Botanic Garden, Sydney, the garden had an operating budget of AU$100 million but every park and garden under the purview of the garden management was declining in the area of funding. The answer was a greater focus on revenue and its generation.[3] Programs that were put in place were provision of a new car park (costing AU$80 million but self-funding), a film studio, and greater use of the golf course, as well as making all events and buildings sustainable from an economic standpoint.[4]

3. Ellis and others (Pollan, 1991) also stress that gardens should deliver social value; this may be as simple as providing space for picnicking or reading or recreation such as yoga or Tai Chi. Whatever the social activity, it is a case of broadening management assets and the management plan. Essentially the garden is a traditional green space but now the garden must be concerned with commercial responsibilities (parking, food, light rail, the disabled) and other assets. As seen in Chapters 7 and 9, this provision of space often takes the form of or has become the venue for large-scale events (Sydney gardens may accommodate upwards of 35,000 plus visitors) but that's the role of the urban botanic garden – being a key factor in creating a livable city, a feature that is rarely enshrined in urban gardens' mission statements.

4. A fourth area in which gardens are singularly placed to contribute in the 21st century is in the long-term intellectual development of society and thus advancing the quality of human life. Gardens need to establish scientific outcomes in the areas of:

a. Physical health.

b. Mental health – and here the work in the Idaho Botanical Garden, Boise may be an example of the first step (see Chapter 8).

c. Education – and here aboriginal traditions have much to offer, particularly in the area of the man–nature divide (see Chapter 6).

d. Outreach to disadvantaged communities.

e. Social justice – and here the work of Michael Pollan and others (e.g. Emmett, 2016) who have explored the link between gardens and social justice becomes important.

These are new areas for gardens to develop, along with the associated funding schemes, to advance community health in the coming years.

5. Chapters 4, 7, and 9 gave only a hint of what might be delivered in innovative programming. Future new directions that could lead to increases in visitation include:

a. Intergenerational programming. This goes beyond the trend in children's gardens or cocktails in the garden and focuses on ways of bringing seniors and youth together in meaningful ways.

b. Reconnection with nature. This can take many forms, from forest bathing to yoga or Tai Chi in the garden to geocaching.

c. Site-specific exhibits. Moving beyond Chihuly-like blockbusters, site-specific exhibits might profile an environmental sculptor, emerging artist, or floral designer.

d. Integration of gardens and performing arts. Lots of gardens bring in big name musical groups, but more are now showcasing dance troupes, international musicians, Shakespearean performances, even acrobatic acts.

e. Biocultural interpretation. This could take the form of anything from international food sampling stations, to apps focusing on the importance of certain plants to particular cultures, to storytelling about cultural histories.

In the case of the Royal Botanic Garden, Sydney, the goal was to engage the public with the goals of the garden while recognizing the history. With 6.5 million visitors, social engagement was paramount and customer service aligned with their values. Thus, programs that engaged the visitor such as apps in Spanish and Mandarin were put in place and then this engagement was continually reevaluated based on the principle the garden is "only as good as the last show".

The remaining four areas of new directions relate principally to the branding and management of gardens:

6. Public demeanor/behavior in a garden that has no admission versus an admission fee. Many public gardens in the UK and USA have no admission charge. It is often the case that the public that comes to a garden for free treats the garden/ collections with less respect than those who had to pay to come in. This has become a more of a trend in the last few years as it was much less of a problem in the past. Today, in the midst of a coronavirus pandemic, it is reasonable to assume from both the public funding perspective and the commercial business perspective that gardens will have to pay their way and that almost certainly will necessitate charging admission.[5]

7. Identity crisis. A vast proportion of garden visitors think of public gardens as a municipal "park", as in the type where one plays baseball or soccer, flies a kite, and the kids climb on jungle gyms, not as a research and education institution, where the plant material has more value than just being "pretty". They do not seem to understand the difference between the garden and its mission (in some cases even mandated by state and national governments[6]) and their local jungle gym or Disney World.

8. In a more general sense those gardens acting as repositories for rare, unusual, or endemic plants[7] also have questions for their future role in society. While it is probably axiomatic that botanic gardens fulfill the conservation and protection role, that is all fine and good; it is useful to some people, but the majority of visitors do not go to botanic gardens to view rare plants or indeed to be educated – and Chapter 6 suggests they do not comprehend it as such, or may not even care about that angle. Most of the visitors who come through a garden do not even look at/ read the signs that explain how the plants might be medicinal or herbal. The question then becomes: If the visitors do not care, then why should gardens bother, in an age when resources for such things are limited? In short, why is it even necessary to be a "reference garden" if no one cares?

9. If the goal is to cater to people on vacation or those people who just want to see something "pretty", and thus marketing is done only to get them in the door and hook them in, then what value is there in creating a place of diverse plant material or significant germplasm? If visitors just want "to see flowers", the question might be asked why, as a horticultural collective, gardens spend exorbitant amounts of time and resources (physical and financial) on unique/ecologically relevant/ (whatever the line is, fill in the blank) gardens when most of the people coming through the gates are just as happy to see rows upon rows, beds upon beds, of petunias, lantanas, palm trees, and azaleas (i.e. Disney World[8])? Indeed there are the select few who actually *understand* horticulture, the point of horticulture, and appreciate the plant diversity in gardens and arboreta, the wild-collected material, etc., but of the numbers that were described in the first edition of *Garden Tourism*, a large percentage of those visitors do not value the *science* part behind the plants. Despite the data saying more people are interested in gardening (in this reference, I mean hardcore gardening, not just visiting a garden for entertainment) than in past years, from my perspective as a gardener of 25 years, that is still a small percentage. What is the cost/benefit analysis of institutions spending many resources on hardcore horticulture versus just creating "pretty" spaces for their visitors' Instagram photos or to sit on the grass for a jazz concert?

10. The transition that was noted above, whereby at the turn of the century gardens went from repositories of plants to having a visitor and marketing business focus, has changed the management perspective. Thus, if one interviewed a garden director in the 1990s, she or he would have espoused that any program held in a garden needed to be mission related. That philosophy or directive has been dismantled under many later directors. If one looks at arboreta or botanic gardens, one can clearly see that they have invited non-mission-related activities/programs into their schedules. Or their mission is not as strict, so they purposely incorporate non-horticultural activities. Thus, one might ask: Have botanic gardens and arboreta turned into a new activity center like the local YMCA? Today one can hear many people say that they feel a botanic garden or arboretum should be the center of community activity, not just horticulture. Are these institutions turning into multipurpose rooms for a city?

The people that are now flowing through garden gates are not old-school gardeners or garden visitors. Times have changed. Where does that leave horticulturists if the new generation of visitors is the Instagram generation that lets their children run through a newly planted garden bed or lets their dog run wild

through a garden? It appears the traditional mission of an arboretum or botanic garden is shifting and garden managers either keep riding a horse or drive a car but either way garden managers must "get on the bus because it's pulling out of the station". But are we, as a society, okay with that? Is the mission of public gardening getting watered down? Do we *have* to water it down lest it die altogether? I am sure this is the perennial question for most tourist facilities like museums, zoos, and botanic gardens. The fear, of course, is that elderly guests and garden clubs are going by the wayside and the young people are not replacing them. The face of public gardening is changing, and what are gardens gaining/losing in the process?

You may have gathered that I am a bit cynical about inviting hordes of visitors into what has been the hallowed grounds and sacred halls of horticulture, aka the public gardens and arboreta of the world. Many have witnessed the shift in public behavior/mentality, and are not sure how to manage the changing personalities we encounter. Furthermore, the potential disparity (or similarity) between a garden administration's management philosophy versus that of its gardeners/workers can be palpable.

Does this approach develop more horticulturists or horticulture-loving people? Again, we are distinguishing between the horticulture-loving versus the "I like pretty flowers" people. The number of official horticulture programs at colleges and universities is nearing zero in the USA. And gardens struggle to find appropriately schooled interns to fill internship positions. Most students are in environmental programs (if we are lucky); few are pursuing horticulture as a career prior to their interning. This is not a wholly terrible thing; many garden managers were biology majors in college and had never heard of horticulture prior to my first internship. But the fact that fewer and fewer young people are entering college with the intent of studying horticulture is still of concern. Maybe other similar cultural institutions have a different experience than we do. But it is a trend worth looking into.

These are all questions gardens are am grappling with. Trying to change one's focus or mentality, even in a national collection, is challenging. Change is sometimes good and sometimes bad.

Case Study 11.1: Where Science Meets Education Meets Conservation Meets Visitors – Gardens as Floral Resources for Pollinating Insects; The Case of the National Botanic Garden of Wales

Seventy-five percent of the world's crops rely on some form of animal pollination, which equates to one-third of all global crop production. Tomatoes, cherries, apples, raspberries, strawberries, coffee, and chocolate all rely on natural pollinators. Research in 2019 suggested that wild bees are disappearing from large swathes of the English countryside (Powney et al., 2019).[9] The situation is similar in the USA and mainland Europe. This decline has been attributable primarily to loss of habitat, particularly ancient wildflower meadows[10] which provide nectar and pollen for the bees, and the introduction of pesticides, particularly neonicotinoid insecticides that were introduced in 2006 but banned in 2018. This decline is of great concern as managed honeybees meet about one-third of Great Britain's pollination needs while wild bees and hoverflies meet the other two-thirds. The pollinating insects are believed to be worth over £700 million per year to farmers. While the number of pollinators of crops has increased it is the more specialized pollinators that are disappearing, and it is the more specialized pollinators that are important for such crops as fruit trees, especially apples, and the wild bees are crucial for the pollination of wildflowers in wildflower meadows. The result is a loss of biodiversity and therefore a dangerous reliance on a small group of pollinators which in turn is dangerous for long-term food security.

Botanic gardens, and especially large botanic gardens, are in a perfect position to test which plants are the most important to pollinators and then disseminate these findings to the gardening public. Such a program is underway at the National Botanic Garden of Wales. With an area of 560 acres (including a working farm!), over 8000 types of plants (in 437 different genera), and a dedicated wild garden, research into bees' plant preferences has become a major research endeavor at the garden. In addition to the above natural landscapes for the bees to use, there is a bee garden with 250,000 bees

in hives in front of an enclosed viewing area. Pollen on the bodies of the pollinators is identified using DNA barcoding to see which plants they have visited.[11] More specifically the garden has also established 20 garden plots within which different seed mixes are cultivated to see which of three types of pollinator (bumblebees, hoverflies, and solitary bees) go to what plants and the season when they go to them. The observational techniques may be expanded to honeybees, butterflies, and moths to see what they prefer. Initial findings suggest that even though the garden has over 8000 species of plants, the bees only use 11% of these; plants in hedgerows and woodlands are preferred and only a small number of garden plants such as dandelions, cotoneaster, spring bulbs, and hellebores are visited.

The goal is to produce pollinator-friendly seed mixes to be used in gardens and amenity areas, assist farmers in managing their land to increase pollinator populations, and with the results from honey pollen supplied by UK farmers, researchers can look at honeybee foraging on a wider scale.

Case Study 11.2: Where are the Future Professionals for Garden Management?

The last student in the UK to enroll in a Botany degree did so in 2010 and graduated from the University of Bristol in 2014. It was predeceased by the degree at the University of Reading which was discontinued in 2007.[12] This decline in botany was mirrored in the USA where, in 1971, 8% of colleges offered a degree in Botany but by 2006 less than 5% offered Botany as a distinct field.[13] While it is still possible to obtain a degree in Plant Science and many students interested in plants can train in the courses offered under Biological Sciences, particularly in the fields of Ecology and Conservation, the fact is that fewer and fewer trained botanists are available to staff the rise in gardens and garden management (see Fig. 11.1). Maybe more worrisome is the decline in plant science education that should be such an important part of fulfilling the education mission of many gardens. In the field of Horticulture approximately 600 students obtain a bachelor's degree in any one year. In contrast, the rise of graduates with a business degree, with 645,000 in 2015, as well as tourism and hospitality[14] with over 10,000 graduates, suggest the paradigm and management shift from botany and plant science toward tourism and hospitality. Thus, for garden management, we might expect to see a greater emphasis on the business of gardens as opposed to the botany of the collection.

In response to the decline of botany as a subject taught in K–12 schools and at the university level, Fairchild Tropical Botanic Garden in Coral Gables, Florida has created a "high school" to teach botany. The program has been placed within the Magnet Schools of America, a public high school but a school of choice (self-selecting with a lottery system to get in) for 500 students. In the four years of the high school, Year 1 contains a Zoology option; in Year 2 students take Botany (one day per week); and in their junior year, Year 3, they choose a major. In 2019 two-thirds of the students chose Botany. In their third year the students spend every second day at Fairchild Tropical Botanic Garden undertaking a research project in the garden. In Year 4, their senior year, they are partnered with a Fairchild scientist prior to going to an advanced education institution. 2019 was the first graduating class. Significantly, 70% were Latino, 20% African American, and 10% white. The program has proven so successful and the demand so high that Botany and Zoology have been infused into the curriculum at an elementary school, Pearce Elementary, with plans to extend this program to other Miami-Dade elementary schools.

Conclusion

All the issues raised above require involvement of governments, gardens, and the public to address these concerns. To conclude the book, following are four management strategies that were suggested by the author, based on the terms of reference in its call for submissions, to the UK Parliamentary Commission on Garden Tourism that met to address some of these issues in 2019. The strategies recommended were:

1. How can gardens across the UK be supported to attract visitors and to ensure their future sustainability?

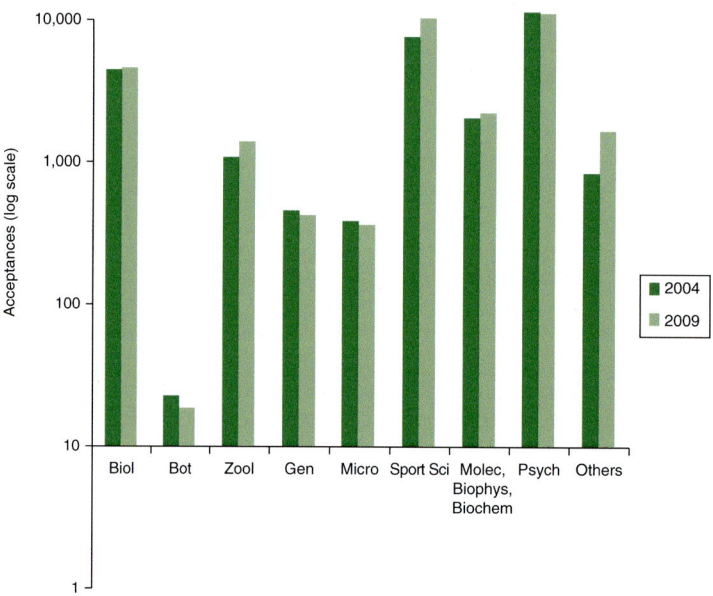

Fig. 11.1. University applications for areas of study in selected biological science subjects in the UK, 2004 and 2009. Biol, Biology; Bot, Botany; Zool, Zoology; Gen, Genetics; Micro, Microbiology; Sport Sci, Sports Science; Molec, Biophys, Biochem, Molecular Biology, Biophysics, Biochemistry; Psych, Psychology.

First, it is time national tourism research organizations recognized garden tourism for its importance and impact. Thus, VisitBritain needs to parse out garden visiting as a separate and distinct tourism activity undertaken by domestic and overseas tourists. Currently garden visiting is subsumed under "Parks and Gardens" and even placed in the "Experiencing City Life" category when much (most?) garden tourism is done outside urban areas! For a simple but dramatic example, the Eden Project in Cornwall, with over a million visitors, would not be included in the garden portion but in city life! Remarkably, VisitBritain shows garden visiting is still the most popular activity in urban tourism (ahead of museums and art galleries!) in the Foresight Report. Furthermore, the segment "Exploring History and Heritage", while this is a popular tourist activity in and of itself, fails to recognize that many of the famous English Heritage and historical sites are gardens! One thinks of Chatsworth, Kenilworth with its new Tudor garden, and most National Trust historical properties have gardens and parks (think: Stourhead, Stowe, Sissinghurst) and research has shown, in some cases, fewer than 10% of visitors go into the historic house, preferring the garden!

The second major suggestion is for VisitBritain to convene a national conference on garden tourism to formulate policy toward gardens and plan new directions. Currently gardens receive little monetary support and even less policy recognition. Now is the time to change that as the Committee has already suggested.

Third, notwithstanding that it has been shown that garden tourism is very important to Great Britain (repeat 33% of overseas tourist want to visit a garden), it is recommended that readers and committee members go on the VisitBritain website: under "Things to do" there is not even a mention of gardens. This is truly remarkable with no explanation of why that is so.

The forgoing represents changes that could be made immediately and thus immediately improve the status of garden tourism in the nation. Other measures of a policy nature that might be considered are:

- Development of a national center for excellence in sustainable tourism.
- Support of, and investment in, skills and ambition development, e.g. degree courses in entrepreneurship and garden management; there are numerous studies charting and lamenting the end of Botany as a discipline in British universities (see above). This should be addressed immediately.
- Focus on targeted destination marketing aimed at new garden visitor markets (i.e. not France and Italy who know our gardens, but such places as South Asia, Australasia, and Africa).
- Adopt models which explore how arts/culture and tourism have aligned and collaborated in recent years – e.g. the Arts Council England (ACE) cultural destinations program – and expand future programs to include collaborations between arts/science/horticulture, with outcomes such as:
  ○ Enabling gardens to work in partnership with destination organizations to increase their reach, engagement, and resilience through working with the tourism sector.
  ○ A contribution to economic growth of gardens and tourism visitor economies and support the development of world-class gardens and tourism products – which meet the needs of domestic and international visitors.
  ○ Development of new audiences, so more diverse audiences are enjoying and exploring gardens and thus contributing to growth of local visitor economies – leading to greater sustainability and resilience for gardens and tourism businesses in local destinations.
  ○ (Re-) Positioning of gardens as a prominent part of the visitor offer and local economic growth plan.
  ○ Commitment from public and private sector partners to continue working in partnership, to support local visitor economy through garden tourism.
- Establishment of a central/national resource/hub for gardens – e.g. the equivalent (or offshoot) of the Association of Leading Visitor Attractions (ALVA) – including supporting smaller regional garden networks; this would offer access to research resources, benchmarking, professional development, and training.

- Support of audience development for gardens through provision of tools and support organizations, e.g.:
  ○ Gardens' equivalent of Audience Finder[14] – to help gardens understand/grow their audiences.
  ○ National group/network for specific audiences – e.g. equivalent of Kids in Museums,[15] where gardens can sign up to a kids' manifesto and engage in nationwide activities.
- Support of/investment in social prescribing – gardens as a source of mental and physical health benefits.
- Support of/investment into a national wildflower center – with the aim of using wildflowers to bring biodiversity, delight, and color into the lives of communities, bridging social divides and stereotypes.

2. What contribution do gardens and garden design make to domestic and international tourism?

The committee was referred to the seminal study by the Oxford Economics (2018) group (see Chapter 8), but it is worth reiterating the importance of garden tourism to the nation.

The Oxford study is the only national study of the impact of garden tourism on a national economy anywhere in the world. One of the key findings was that spending on ornamental horticulture and landscaping contributed over £24.2 billion in 2017. Of that figure, £2.2 billion could be attributable to overseas tourists visiting gardens and parks or one-fifth of all tourism spending by overseas tourists. To this figure should be added £690 million in spending by domestic tourists including both overnight domestic travel and day trips. In total, direct visitor spending on parks and gardens contributes £1.2 billion to the nation's GDP and was associated with 32,000 jobs and £265 million in tax revenues. Adding indirect and induced contributions suggests gardens add £24.2 billion in GDP and £5.4 billion in revenues to the government. These figures are comparable to other regions of the world where garden tourism is a significant economic activity. Thus, for example in the USA, gardening revenues are higher than those of gaming, sports activities and even *Avatar* – the highest grossing revenue producer in movie history!

3. How do gardens and garden design contribute to the creative economy?

Gardens are increasingly one of the few public open spaces where the creative economy can find a vehicle and a voice. There are numerous examples both in this book and elsewhere where gardens and garden managers from such places as Kew, Eden, Chatsworth, and many National Trust properties have dedicated time and space to art installations, the performing arts, recitals, and cultural vehicles for the promotion of art, culture, and diversity. Thus at the time of the parliamentary enquiry, Kew was about to open a Chihuly glass exhibit, Eden to host Kylie Minogue, and even at the local level, Berrington Hall in Shropshire, the garden management had positioned, in the center of the historic walled garden, "LOOK!", a 20 ft installation that offers a space in which to take in the natural surroundings, enjoy the gardens, and throughout the summer visitors can enjoy live music and talks inside it, which is exactly how pavilions such as these were used in the eighteenth century.

4. How can garden design and landscaping best support community spaces and community connections?

This is perhaps the one area where gardens have seen an increasing role and responsibility in their outreach beyond the garden walls. These outreach activities often are draws and vehicles by which the domestic and international tourism community can see Britain. The best example may be in the heart of London where community outreach and connections from their gardens is done by the City of Westminster. Westminster City Council is responsible for 54 community and local gardens, parks, and cemeteries whereby they provide green space in an urban area for both local communities and urban workers. To develop such outreach activities it is important they receive financial and logistical support to expand such a valuable part of the urban landscape and project a world-class garden organization; by doing so it is readily apparent they also act to drive visitors to gardens and parks throughout the country. On any fine day in London while walking along to Whitehall Gardens one can see large numbers of visitors and workers, or close by the small but intimate and beautiful Drury Lane Gardens or even Victoria Tower Gardens; all are testament to what value garden design and management can do to promote outstanding tourism facilities and hospitality events.

Notes

[1] A refresher by reading *The Culture of Flowers* by Jack Goody (1993) is suggested here.

[2] K. Ellis, Sydney, 2018, personal communication.

[3] This, of course, is immediately criticized as selling out the garden's mission, creating a commercial garden, and typical of neoliberal emphasis on capital accumulation rather than social need (see Chapter 9 and Smith, 2019).

[4] The garden stressed that there was no point in doing an event if it spoils the green space or does not use existing buildings, and thus physical sustainability was also a major consideration.

[5] The two examples where this may not apply is Australia and New Zealand, both with very few coronavirus fatalities (Australia had fewer than 100 fatalities from the pandemic in mid-May).

[6] In the case of some gardens in Washington, DC and to a certain extent Kew Gardens in London, their mission is mandated by the federal/national government.

[7] The best example of these types of garden are reference collections for the national plant societies. Thus, the US National Arboretum in Washington, DC is home to the reference collection of the National Herb Garden, while America's Rose Garden in Shreveport, Louisiana is the home of the National Rose Garden and its society.

[8] Walt Disney was often quoted as saying: "I would rather entertain and hope that people learned something than educate people and hope they were entertained."

[9] The research shows that each square kilometer lost an average of 11 bee and hoverfly species between 1980 and 2013; the range of the remaining bees declined at least 25% in those years and in some upland regions the decline may be as high as 50%.

[10] In the UK 97% of wildflower meadows were lost between 1930 and 2016.

[11] The National Botanic Garden of Wales is a leader in this field. It made Wales the first nation in the world to have a reference library of DNA barcoding of all its native plants (see De Vere *et al.*, 2017).

[12] The number of garden managers who graduated from Reading over the years was remarkable.

[13] In contrast, in 2009, the discipline of Zoology in the UK had an enrollment of 1400 students, with 15,000 taking Psychology and 19 Botany students.

[14] Data are grouped to include hotel management, tour and travel, meeting and event management, and casino management. For detailed a breakdown, see https://nces.ed.gov/programs/digest/d13/tables/dt13_318.30.asp (accessed July 29, 2020).

[14] Available at: https://audiencefinder.org/about/ (accessed July 31, 2020).

[15] Available at: https://kidsinmuseums.org.uk/who-we-are/about-us/ (accessed July 31, 2020).

Glenstone Museum; where art meets flora. Jeff Koons Split Rocker, a 36 foot tall living structure. Photo courtesy of Glenstone Museum.

Fairchild Botanic Garden in Coral Gables, Florida has an active research program with NASA to integrate flora into the space program both for pure research and for plant growth in space for food.

Plant extinction is a major concern of the IUCN. Kew is a major research institution preventing extinction – note the reference to the IUCN in the signage board.

… and while many plants are in danger of extinction there are many yet to be discovered. Café marron from Mauritius was one such species but, in this case, rediscovered.

The discipline of Botany is disappearing from the world's universities and institutions of higher education. The young lady horticulturalist working in the garden is also becoming a rare species.

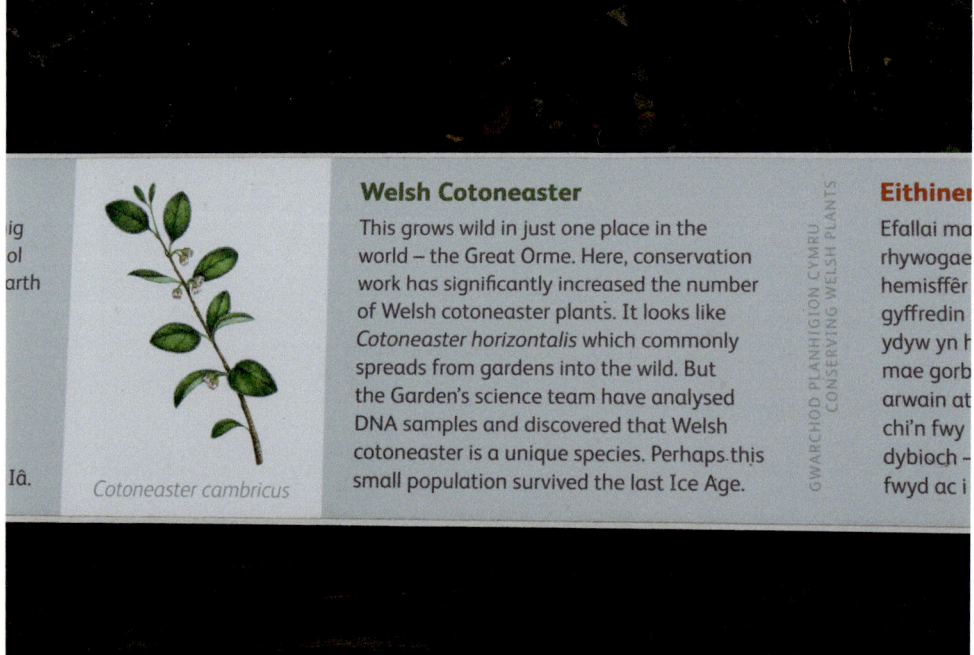

Welsh Cotoneaster

This grows wild in just one place in the world – the Great Orme. Here, conservation work has significantly increased the number of Welsh cotoneaster plants. It looks like *Cotoneaster horizontalis* which commonly spreads from gardens into the wild. But the Garden's science team have analysed DNA samples and discovered that Welsh cotoneaster is a unique species. Perhaps this small population survived the last Ice Age.

Cotoneaster cambricus

GWARCHOD PLANHIGION CYMRU
CONSERVING WELSH PLANTS

Eithinen

Efallai ma
rhywogae
hemisffêr
gyffredin
ydyw yn h
mae gorb
arwain at
chi'n fwy
dybioch –
fwyd ac i

This image illustrates how the National Botanic Garden of Wales, Carmarthenshire, fulfills its role in saving endangered plant species. Author's own photo.

The Canadian Tulip Festival, held in Ottawa, Ontario and started in 1953 from tulip bulbs gifted in gratitude by the Dutch government for the sheltering of the Dutch Royal Family during World War II, today displays over 3 million tulips annually and receives over 650,000 visitors. Photo courtesy of Canadian Tulip Festival and used with permission.

12

The Effect of the Coronavirus on Garden Tourism[1]

© Richard W. Benfield 2021. *New Directions in Garden Tourism* (R. Benfield)
DOI: 10.1079/9781789241761.0012

In January 2020, the Covid-19 virus first made its appearance in North America.[2] In February the warning of a pandemic was first raised[3] and by March the world was in the throes of a fully developed pandemic. To counter the virus, on March 13, US authorities announced/declared a national emergency and within 48 hours gardens across the USA began to close. Within 48 hours, 17% of US public gardens were closed and by March 16, following the release of national social distancing guidelines, 26% of gardens had closed and 15% closed the day after. By March 30, only 4% of APGA member gardens were "fully open" and 60% were fully closed,[4] and this pattern of garden closing had been repeated throughout the world.

Effects of Gardens Closing

The immediate effect was of course the lack of earned income from admissions, but the repercussions were wider:

- 91% canceled education classes;
- 83% canceled external events (weddings and external meetings);
- 69% canceled ticketed events;
- 64% canceled their own fundraising events; and
- membership sales fell by 39%.[5]

Estimates show that, collectively, member gardens of the APGA were losing over US$3 million per day in earned income[6] – for an average size garden this would be almost US$190,000 per month.

Staffing impacts were almost as immediate. Fifty-one percent began working from home,[7] 30% reported layoffs or furloughs, and concern about the effect on the valuable plant collections was widespread. In April, the situation worsened ...

April 2020

Almost all gardens remained closed during the month[8] of April. This was because over 75% had been mandated to close by their parent institution although many had justified closure owing to "staff concerns". Even though many gardens were closed they donated personal protective equipment to hospitals and sent flowers from the garden to nursery homes and hospitals. In many cases virtual tours and virtual education were provided on the Internet, along with updated images from gardens that, as it was spring, would have been at their height of popularity. Staffing impacts continued to be severe as 40% reduced staff hours and 37% were on rotating schedules. Admirably only 27% had to resort to layoffs or furloughs although the impact of increased workload on a smaller staff was often an issue. This was particularly a problem as most gardens (77%) suspended their on-site volunteer contribution.[9] The result of the continued closure was serious both from a cash flow perspective as well as from long-term budgetary needs and expenditures.[10] The financial future was made even more grim by a fall in memberships (57% noted this fall) and 65% of gardens had canceled their own fundraising events. This was even more worrisome as, on average, 27% of a garden's expenses (around US$42 million per month) go to horticulture and facilities, those areas that protect the valuable plant collections.

The two federal government programs that were passed to alleviate the stress on US businesses[11] unfortunately could not be accessed by many gardens owing to them being a subsidiary part of a larger parent organization (universities and colleges). Of those who could and did apply for assistance, only 21% received approval for money. Other grant sources were available, but few appear to have accessed them. Concurrently donations and private funding, that also would have helped opening, had diminished and hence were not available to help in the recovery.

May 2020

By May 20, all states in the USA had "opened" but the opening of gardens was variable across the country. California Governor, Gavin Newsom, specifically mentioned botanic gardens in his Phase 1 reopening approvals and as early as May 4 some gardens were doing limited openings (so-called "soft openings")[12] but with strict measures in place to manage visitors (see below). In contrast, following May 20,[13] few gardens on the eastern seaboard were open and had no idea when they would be permitted to open.

Management of Gardens under the Coronavirus

Most gardens in the USA and the UK adopted a phased approach to opening the gardens (and associated facilities) with a gradual movement toward fully opening by the final phase. Fig. 12.1 shows the phased opening plan of Mt. Cuba Center,[14] Hockessin, Delaware, that reopened on June 17, 2020.

Most of Phase 1 of opening gardens involved passive viewing without food services, retail openings, events, and often washroom facilities. For visitation, the major issue in reopening gardens after the coronavirus was the need to respect social distancing or keeping individuals/households at least 6 ft apart. Such requirements meant that gardens had to enact two major management strategies. The first was to keep people apart by timing the entry. This was accomplished using timed, pre-reserved[15] entry tickets (Fig. 12.2). This was essentially management at the front gate and, from there, staggering the entry. Based on the patterns of movement of visitors through the garden, the second requirement was to establish what the carrying capacity of the garden was and specifically the areas where visitors congregate. This second stage was invariably done through historical knowledge of garden numbers, the patterns of visitation[16] and movement, and from there guests could be managed by means of one-way traffic and docents moving the guests through the garden (see Figs 12.3 and 12.4 below).

The Future of Garden Tourism after the Coronavirus

Because of the dramatic diminution in international airline services (as well as domestic services), it is expected that the summer of 2020 will be marked by short-haul domestic road travel both in the USA and Europe.[17] The UNWTO is forecasting a 20% drop in tourism receipts for the year 2020 or essentially a 7-year drop in receipts from historic norms. The prognosis beyond 2020 would suggest a slow recovery due almost entirely to fear of flying in a confined space. Thus, the future of botanic gardens in all

countries should be considered bullish as long as the revenues are sufficient to meet expenses. To that end, and in conjunction with restricted numbers, one might expect a rise in garden entrance fees. This seems especially important as one can expect large gatherings at events and concerts to be the last phase of gardens opening.

Gardens Outside the USA; Situation in the UK

For gardens in the UK, the major gardens affected were either private gardens opened under the National Garden Scheme or the gardens of the National Trust. For the National Garden Scheme, 2019 was a record year, for the first time ever the gardens raised more than £4 million – an amazing achievement. That meant they were due to make record donations to their nursing and health beneficiaries of £3.3 million. That money would have been given out in two tranches in April and July 2020 – the cash pattern is always that funds raised in one year supply the donations in the next, so that by the time the donations are made money is flowing in again from the gardens opening. Simple but effective – until the gardens stop opening, that is, which they have never done before.

As in the USA, gardens closed in mid-March and as the gardens account for 90% of all income, the National Garden Scheme faced a challenge. It worked out a worst-case scenario with no garden income for the rest of the year, then saw how much cash from the donation "pot" needed to be retained, and then gave out very reduced donations. A painful decision.

At the time of writing, the National Garden Scheme has put some cost savings in place but, more important, is running a campaign "Help Support Our Nurses" that focuses on the nursing charities the Scheme supports and the fact they are all on the front line of Covid-19. The campaign is being driven by short videos the garden owners are making of their gardens – Virtual Garden Visits (as opposed to real-life visits), which people are watching and then making a donation. Every week on a Thursday the conservancy puts out a new batch of films and the circulation is worldwide, with a donation in April from as far away as Hawaii!

MT.CUBA CENTER

Gardening on a higher level

Mt. Cuba Center COVID-19 Phased Re-Opening Plan

	Phase 1	Phase 2	Phase 3
Ticketing	Online ticket sales will be highly recommended both in advance and onsite. If necessary for guest resolutions, staff may sell tickets onsite in the guest parking lot, with social distancing measures in place. A self-service scanning station will be located in the guest parking lot for those who have purchased tickets in advance, with a staff member nearby to provide support. A sign with the latest social distancing recommendations in the gardens will be prominently displayed at the guest entrance. Ticketing equipment will be disinfected at least once daily. Self-scanning ticket stations for guest will be sanitized often throughout the day	Online ticket sales will be highly recommended both in advance and onsite. If necessary for guest resolutions, staff may sell tickets onsite in the guest parking lot, with social distancing measures in place. A self-service scanning station will be located in the guest parking lot for those who have purchased tickets in advance, with a staff member nearby to provide support. A sign with the latest social distancing recommendations in the gardens will be prominently displayed at the guest entrance. Ticketing equipment will be disinfected at least once daily. Self-scanning ticket stations for guest will be sanitized often throughout the day	Online ticket sales will be highly recommended. A staff member will sell tickets onsite in the guest parking lot (on busy days) and Front Desk (on slow days). A sign with the latest social distancing recommendations in the gardens will be prominently displayed at the guest entrance. Ticketing equipment will be disinfected at least once daily
General admission capacity management	Members only will be granted access for the first week. After that, General Admission will be open to all, but capacity will be actively monitored by the Supervisor On Duty. If attendance reaches a point where social distancing could become difficult, the Supervisor On Duty will close entry until numbers dissipate, the	Capacities will be actively monitored by Supervisor On Duty. If attendance reaches a point where social distancing could become difficult, the Supervisor On Duty will close entry until numbers dissipate	Capacities will continue to be monitored to maintain the most up to date physical distancing protocols
Membership sales	Online membership sales will be highly encouraged both in advance and onsite. If necessary, a staff member may sell onsite in the guest parking lot, with social distancing measures in place. New membership fulfillment materials will be mailed rather than provided onsite. Fulfillment may be offered on demand when necessary for use of complimentary tickets or gifted memberships	Online membership sales will be highly encouraged both in advance and onsite. If necessary, a staff member may sell onsite in the guest parking lot, with social distancing measures in place. New membership fulfillment materials will be mailed rather than provided onsite. Fulfillment may be offered on demand when necessary for use of complimentary tickets or gifted memberships	Online membership sales will be highly encouraged. A staff member can sell and fulfill memberships onsite, with social distancing measures in place. Physical card can be printed and fulfillment materials distributed at the time of sale
Public programs	Classes will be offered in an online format when feasible. Onsite gatherings (if any) will be limited to 10 individuals. Special events, such as Twilight on the Terrace, will be cancelled	Onsite gatherings will be limited to 50 individuals (including staff). Special events will be cancelled. Garden Highlight Walks and Garden Enthusiast Tours may be offered for sale online and in the guest parking lot, with social distancing measures in place. Any course materials will be shared digitally	Onsite public events will resume in accordance with the most up to date government safety guidelines
Off-site outreach programs	Offsite programs or outreach efforts involving non-essential travel and/or gatherings larger than 10 will be cancelled	Offsite programs or outreach efforts involving groups fewer than 50 will be evaluated on a case-by-case basis to determine if they can be run safely and in accordance with the most recent government guidelines and social distancing measures	Offsite programs or outreach efforts will be evaluated on a case-by-case basis to determine if they can be run safely and in accordance with the most recent government guidelines and social distancing measures
Facilities	The Main House will be closed to guests with the exception of access to the first floor restrooms through the accessible entrance. A CDC-issued sign about how to slow the spread of COVID-19 will be displayed near the entrance, and within restrooms. Facilities will be cleaned and disinfected regularly according to CDC guidelines. Accessible parking will be relocated to the guest parking lot	The Main House will be closed to guests with the exception of access to the first floor restrooms through the accessible entrance. A CDC-issued sign about how to slow the spread of COVID-19 will be displayed near the entrance, and within restrooms. Facilities will be cleaned and disinfected regularly according to CDC guidelines. Accessible parking will be relocated to the guest parking lot	The Main House will be opened to guests with reduced seating to allow for physical distancing guidelines
Public access	Guest circulation will follow a one-way route through the gardens. This will be reinforced by a guest map (delivered via self-service pick-up, or which accessed in advance via MCC's website), signage, and staff present onsite. Social distancing measures will be reinforced by staff present onsite and signage	Guest circulation will follow a one-way route through the gardens. This will be reinforced by a guest map (delivered via self-service pick-up, or which accessed in advance via MCC's website), signage, and staff present onsite. Social distancing measures will be reinforced by staff present onsite and signage	Standard guest circulation will be re-introduced. The appropriate physical distancing measures will be communicated via staff and signage, and monitored by staff onsite
Personal protective equipment	Staff and guests will be required to comply with any government mandates regarding personal protective equipment. MCC will maintain an inventory of face masks, gloves, and hand sanitizer available for staff and guests	Staff and guests will be required to comply with any government mandates regarding personal protective equipment. MCC will maintain an inventory of face masks, gloves, and hand sanitizer available for staff and guests	Staff and guests will be required to comply with any government mandates regarding personal protective equipment. MCC will maintain an inventory of face masks, gloves, and hand sanitizer available for staff and guests
Staff	Select staff will be encouraged to continue to work from home or use emergency leave, including vulnerable individuals. Staff reporting to work will be required to comply with the latest governmental social distancing recommendations	Select staff will be encouraged to continue to work from home or use emergency leave, including vulnerable individuals. Staff reporting to work will be required to comply with the latest governmental social distancing recommendations	MCC will make accommodations as necessary in order to assist vulnerable individuals in returning to work safely. Staff reporting to work will be required to comply with the latest governmental social distancing recommendations
Volunteers	Volunteers may be permitted to return only for tasks that allow for compliance with the latest governmental social distancing recommendations. We will not be hosting outside volunteer groups during this phase	Volunteers may be permitted to return only for tasks that allow for compliance with the latest governmental social distancing recommendations. We will consider hosting outside volunteer groups, of smaller than 20, on a case by case basis	Volunteers will be permitted to resume normal activities. We will encourage appropriate social distancing for all, especially with vulnerable populations. Outside volunteer groups will be invited to return

4/30/2020

Fig. 12.1. Covid-19 Phased Re-opening Plan at Mt. Cuba Center, Hockessin, Delaware, USA. MCC, Mt. Cuba Center; CDC, Centers for Disease Control and Prevention. Plan courtesy of Mt. Cuba Center, Pennsylvania and APGA, and used with kind permission. (Full portfolio of maps available at: https://www.publicgardens.org/resources/mt-cuba-center%E2%80%99s-covid-19-reopening-playbook-employee-safety-and-health, accessed August 14, 2020).

Booking tickets

Tickets are only available for purchase online at designated times to manage capacity and to limit contact upon entry.

[Book now ☑]   [Book as a member ☑]

ⓘ We are not allowing visitors to use any passes for garden entry.

A service fee of $1 will be added to each ticket for non-members. This fee allows us to offer ticketing from an online platform to ensure that we can manage capacity and provide our visitors with the safest, touch-free experience.

Visiting the garden

Visitors are encouraged to arrive on time. There will be a 30-minute grace period.

The recommended visit time is between 30 and 90 minutes.

We encourage you to present your ticket on a mobile device, but you can print it out too.

Affected activities and events

- Parking lot: open for visitors. If you are not able to maintain a 2-metre distance from others when exiting your vehicle, wait inside until it is clear to get out.
- Events, meetings, educational programs, and guided tours: cancelled until end of June
- Rental viewings, elopements and small weddings, engagement photos, or other photo shoots: not permitted
- Maze: open (one person or household at a time)
- Select features and areas (including large lawn chairs and tractor): closed
- Pathways: some will be one-way
- Gift shop and Truffles Cafe: closed

Fig. 12.2. Timed entry tickets at VanDusen Botanical Garden, Vancouver, British Columbia, Canada. Reproduced with kind permission of VanDusen Botanical Garden and APGA. (Available at: https://vancouver.ca/parks-recreation-culture/re-opening-during-covid-19-coronavirus.aspx#book, accessed August 19, 2020).

At the National Trust, the major imperative as its properties are open year-round was the need to maintain social distancing, and to avoid creating hotspots and discourage gatherings. The Trust knew this meant that when the properties reopened, visitor numbers would need to be carefully managed. And previous management of overtourism suggested that the safest way to do this was through an online pre-booking system. This was new for the Trust[18] and it had to work fast to get it in place.

Many Trust staff were furloughed[19] and the gardens were "put to bed", so it took a little while to wake them all up again. The Trust prioritized opening the gates and offering beautiful outdoor spaces first, and always in line with government guidance in England, Wales, and Northern Ireland. From Wednesday May 13, the Trust started opening some car parks in England, so people could access fresh air, open space, and nature. Car parks which were staffed and had facilities, took longer. All car parks in Wales and Northern Ireland remained closed. Members could park for free at the reopened car parks. Interestingly they were also finding that without the usual visitors, wildlife had moved into some car parks and open spaces, so human guests had to be on the lookout for unexpected fellow guests in some places when they returned!

Cafés, shops, gardens, and houses will open in time, but only where and when it is safe[20].

Case Study 12.1: Longwood Gardens, Kennett Square, Pennsylvania, USA, after Covid-19

For many years the tourism literature has been consumed by the concept of *carrying capacity* or how many tourists a specific location can accommodate without exceeding a physical and/or social limit, after which there is impairment of satisfaction, irrevocable physical harm, and essentially overuse, now often couched as "overtourism". Close examination of the literature would show much work on what carrying capacity is, why limits need to be established, how they might be established, marketing strategies to dissuade use, and the results of exceeding carrying capacity – but there are few to almost no studies on empirical numbers of the absolute carrying capacity of a tourist facility. As noted above, establishing carrying capacity to permit safe numbers of visitors to gardens was an absolute priority and necessity for gardens to undertake limited reopening and not surprisingly the industry has risen to this challenge with academics still grappling with the concept. The best example is the most famous (and possibly most heavily visited) garden in the USA: Longwood Gardens. As an example of the second stage of reopening noted above, Longwood Gardens not only set carrying capacity limits for individual areas of the garden (Fig. 12.3) but also set patterns of movement through the garden and specifically in the Conservatory[21] (Fig. 12.4), thus securing and enhancing social distancing.

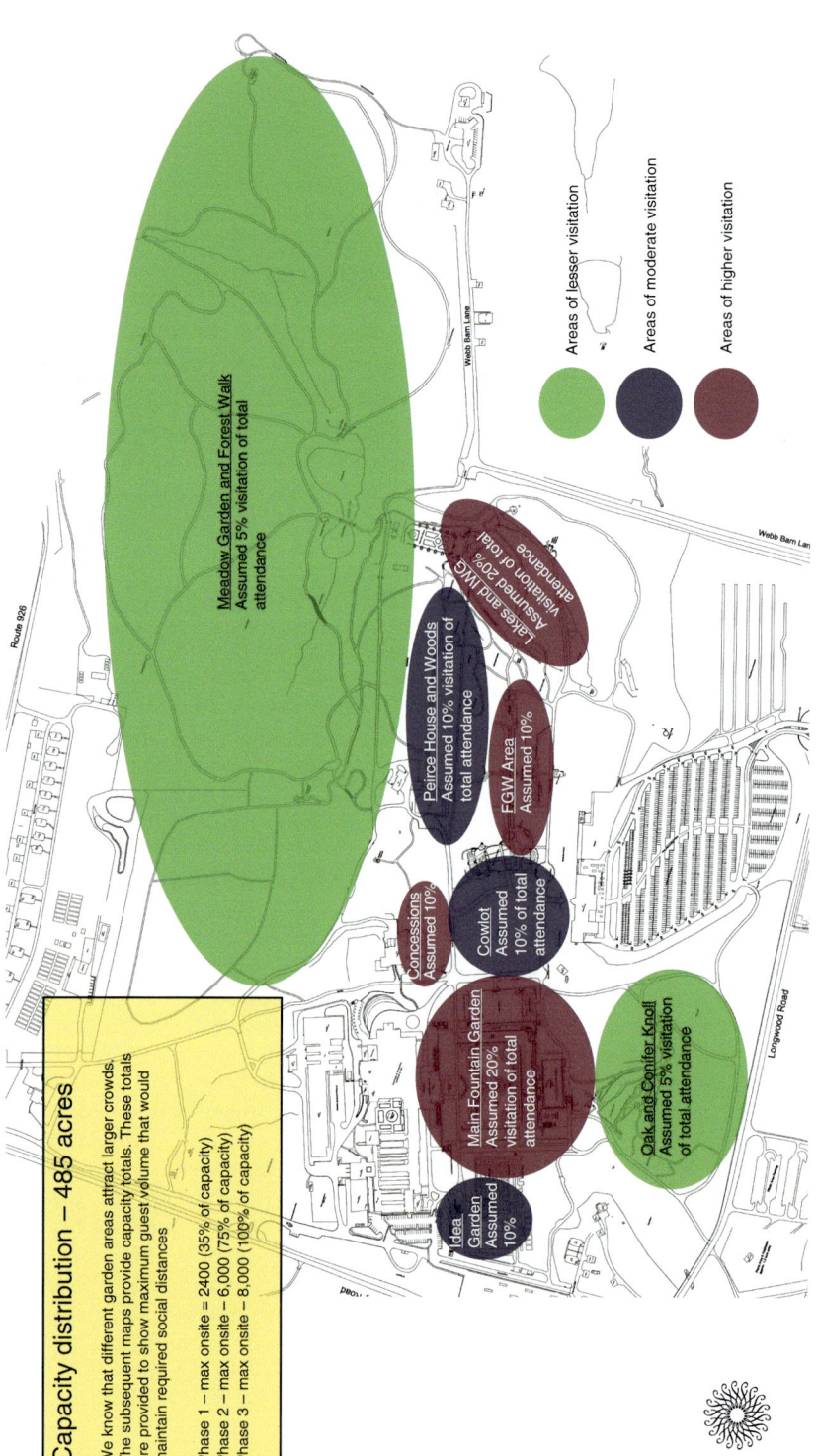

Capacity distribution – 485 acres

We know that different garden areas attract larger crowds. The subsequent maps provide capacity totals. These totals are provided to show maximum guest volume that would maintain required social distances

Phase 1 – max onsite = 2400 (35% of capacity)
Phase 2 – max onsite – 6,000 (75% of capacity)
Phase 3 – max onsite – 8,000 (100% of capacity)

Meadow Garden and Forest Walk
Assumed 5% visitation of total attendance

Lakes and IWG
Assumed 20%
visitation of total
attendance

Peirce House and Woods
Assumed 10% visitation of total attendance

FGW Area
Assumed 10%

Concessions
Assumed 10%

Cowlot
Assumed 10% of total attendance

Main Fountain Garden
Assumed 20% visitation of total attendance

Idea Garden
Assumed 10%

Oak and Conifer Knoll
Assumed 5% visitation of total attendance

Route 926

Webb Barn Lane

Webb Barn Lane

Longwood Road

Areas of lesser visitation

Areas of moderate visitation

Areas of higher visitation

Fig. 12.3. Capacity distribution at Longwood Gardens, Kennett Square, Pennsylvania, USA. IWG, Italian Water Garden; FGW, Flower Garden Walk. Map courtesy of Longwood Gardens, Pennsylvania and APGA, and used with kind permission. (Full portfolio of maps available at: https://www.publicgardens.org/resources/longwood-capacity-distribution-and-guest-circulation-covid-19-reopening-plan, accessed August 14, 2020).

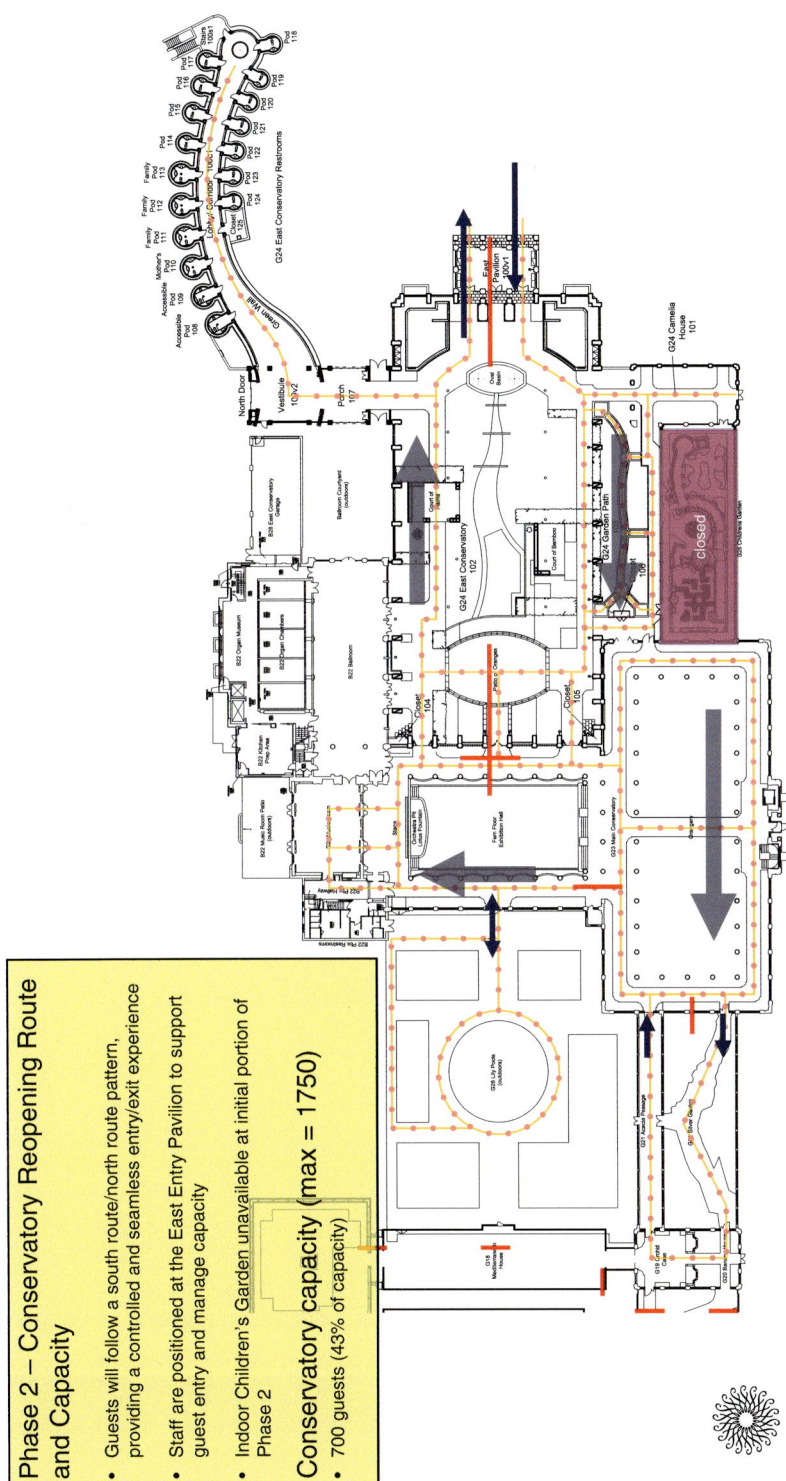

Fig. 12.4. Conservatory reopening route and capacity at Longwood Gardens, Kennett Square, Pennsylvania, USA. Map courtesy of Longwood Gardens, Pennsylvania and APGA, and used with kind permission. (Full portfolio of maps available at: https://www.publicgardens.org/resources/longwood-capacity-distribution-and-guest-circulation-covid-19-reopening-plan, accessed August 14, 2020).

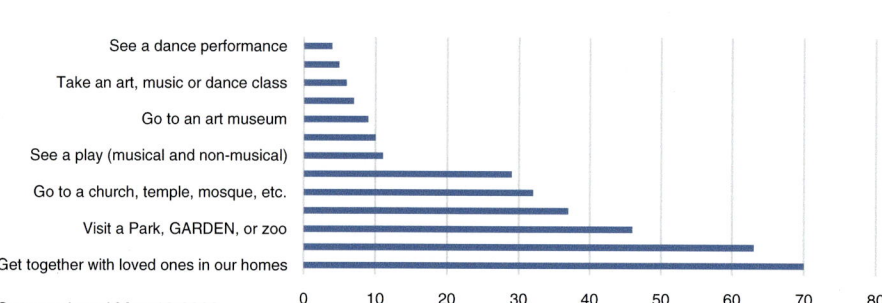

Thinking Ahead when people are able to go out again, what are you most excited to do in the first few weeks (could check up to five)

Survey closed May 19,2020

Fig. 12.5. Prospects for Visitation at US Botanic Gardens when COVID restrictions lifted. Courtesy of Culture Track/LaPlaca Cohen.

As this book goes to press, most gardens in the USA are open but with limited or non-existent facilities for visitors. The case is very much the same in the UK, including a limited number of gardens open as part of the National Garden Scheme. What is heartening is that when the Covid-19 pandemic is over gardens will be the most popular tourism visitation option (Fig. 12.5), as shown by a study by Culture Track/LaPlaca Cohen[22], and this surely is the best ending to a garden tourism book ... ever.

Notes

[1] The majority of the data presented herein were made available courtesy of the APGA who conducted extensive surveys in March and April 2020. The author is grateful for permitting use of these data.

[2] January 21, 2020.

[3] The World Health Organization expressed concern on January 31, the disease was given a name, Covid-19, on February 11, and a pandemic was confirmed on March 11, 2020.

[4] Thirty-six percent were partially open but most of those soon became fully closed.

[5] Although those partially open gardens saw an increase in membership sales.

[6] This figure on a normal year would represent 63% of income.

[7] Forty-five percent of gardens retained horticultural staff.

[8] And 11% went from partially to fully closed (71%).

[9] This was a problem because volunteers provide, on average, 1 hour of free labor for every 6 hours worked by staff.

[10] Fifty-six percent of gardens cancelled ALL programs.

[11] The CARE Act (March 27) and the HEROES Act (May 15).

[12] The garden at Filoli Historic House & Garden, on this book's cover, opened to members only on May 4 and with highly regulated pathways and social distancing.

[13] May 20 was the first day all fifty states were "open" for some form of commercial activity.

[14] That in July 2020 was voted top botanical garden in *USA Today*'s 10 Best Readers' Choice Awards (see Chapter 5).

[15] Which had the additional benefit of mandating online reservations and hence online payment, thus obviating the need for cash transactions and handling.

[16] In this regard the GPS research conducted in Chapter 6 and what it could provide would have been invaluable.

[17] There is anecdotal evidence that recreational vehicle (RV) dealers in the USA saw a major growth of RV rentals and even sales in March, April, and May of 2020.

[18] While the Trust suggests that online booking was a new venture, it had used similar management measures at major gardens in the past (see Benfield, 2001).

[19] At the Royal Botanic Gardens, Kew (NOT a National Trust property), 60% of staff were furloughed with the remaining 40% being gardeners and senior management.

[20] Most cafés and retail outlets opened on July 4, 2020. Houses opened on the same day, but often with reduced rooms to visit, e.g. Snowshill Manor, Gloucestershire, UK.

[21] Notwithstanding the Conservatory at Longwood Gardens is the largest in the nation.

[22] Some unpublished studies suggest this percentage may be as high as 60%.

Booking tickets

Tickets are only available for purchase online at designated times to manage capacity and to limit contact upon entry.

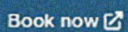

ⓘ We are not allowing visitors to use any passes for garden entry.

A service fee of $1 will be added to each ticket for non-members. This fee allows us to offer ticketing from an online platform to ensure that we can manage capacity and provide our visitors with the safest, touch-free experience.

Visiting the garden

Visitors are encouraged to arrive on time. There will be a 30-minute grace period.

The recommended visit time is between 30 and 90 minutes.

We encourage you to present your ticket on a mobile device, but you can print it out too.

Affected activities and events

- Parking lot: open for visitors. If you are not able to maintain a 2-metre distance from others when exiting your vehicle, wait inside until it is clear to get out.
- Events, meetings, educational programs, and guided tours: cancelled until end of June
- Rental viewings, elopements and small weddings, engagement photos, or other photo shoots: not permitted
- Maze: open (one person or household at a time)
- Select features and areas (including large lawn chairs and tractor): closed
- Pathways: some will be one-way
- Gift shop and Truffles Cafe: closed

Timed ticket and entry conditions, Van Dusen Botanic Garden, Vancouver, British Columbia, Canada. Image courtesy of American Public Gardens Association, Pennsylvania.

At Bourton House Gardens, like most gardens, staff were furloughed during the early months of the pandemic and only returned to work six or eight weeks after being laid off. Tom Benfield (no relation!), Assistant Gardener at Bourton House, was furloughed for six weeks and faced significant deferred maintenance upon his return.

Chelsea Flower Show goes remote during the 2020 Covid crisis. Photo courtesy of Royal Horticultural Society.

Covid-19 has necessitated one-way systems of movement throughout most gardens to promote safety through social distancing of visitors. This image is from Bourton House Garden, Gloucestershire. Author's own photo.

At most gardens, lines for usually pre-paid entry, are socially distanced at intervals of six feet. At Alnwick Garden, Northumberland, shown here, masks are mandatory and all transactions, such as payments, are remote. Author's own photo.

Selected Bibliography

Aaker, J.L. (1997) Dimensions of brand personality. *Journal of Marketing Research* 34(3), 347–356.

Acharya Jagadish Chandra Bose Indian Botanic Garden (Howrah, India). https://tools.bgci.org/garden.php?id=53 (accessed August 10, 2020).

Aga Khan Garden (Edmonton, Alberta, Canada). https://botanicgarden.ualberta.ca/aga-khan-garden-alberta/ (accessed August 10, 2020).

Alexander, A. (2018) Hulshosch repositions Australia's oldest Botanic Gardens. *B&T Magazine*. Available at: https://www.bandt.com.au/hulsbosch-repositions-australias-oldest-botanic-gardens/ (accessed July 20, 2020).

Allen Centennial Garden (Madison, Wisconsin). https://allencentennialgarden.wisc.edu/ (accessed August 10, 2020).

The Alnwick Garden (Alnwick, UK). https://www.alnwickgarden.com/ (accessed August 10, 2020).

American Academy of Arts & Sciences (2019) Historic Sites Visits. Available at: https://www.amacad.org/humanities-indicators/public-life/historic-site-visits (accessed August 14, 2020).

American Marketing Association (AMA). Available at: https://www.ama.org (accessed July 20, 2020).

American Public Gardens Association (APGA). Available at: http://www.publicgardens.org/ (accessed July 30, 2020).

American Public Gardens Association (2015) American Public Gardens Association unveils brand refresh. Available at: https://www.publicgardens.org/news/article/american-public-gardens-association-unveils-brand-refresh (accessed July 20, 2020).

American Public Gardens Association (undated) What is a public garden? Available at: https://publicgardens.org/about-public-gardens/what-public-garden (accessed July 20, 2020).

America's Rose Garden (Shreveport, Louisiana). https://www.rose.org/visit-public-gardens (accessed August 10, 2020).

Arcadis Consultants (2016) Royal Botanic Garden Edinburgh: Economic Impact Assessment. Available at: https://www.rbge.org.uk/media/5468/economic-impact-assessment-oct-2016.pdf (accessed July 30, 2020).

Arizona-Sonora Desert Museum (Tuscon, Arizona). https://www.desertmuseum.org/ (accessed August 10, 2020).

Association of Leading Visitor Attractions (ALVA). https://www.alva.org.uk/index.cfm (accessed August 19, 2020).

Association of Mexican Botanic Gardens (AMJB). Available at: https://www.concyteq.edu.mx/amjb/ (accessed 3 November 2020).

Atlanta Botanical Garden (Atlanta, Georgia). Available at: https://atlantabg.org/ (accessed July 20, 2020).

Auckland Botanic Gardens (Auckland, New Zealand). http://www.aucklandbotanicgardens.co.nz/ (accessed August 10, 2020).

Australian Arid Lands Botanic Garden (AALBG) (Post Augusta, South Australia, Australia). https://www.aalbg.org/ (accessed August 10, 2020).

Bacon's Castle (Surry, Virginia). https://preservationvirginia.org/historic-sites/bacons-castle/ (accessed August 10, 2020).

Baldwin, Ian (2018) Opinion: Be aware of the under-35 demographic. Garden Centre News. Available at https://www.gardencentermag.com/article/ian-baldwin-35-demographic-opinion/ (accessed October 15, 2020).

Ballantyne, R., Packer, J. and Hughes, K. (2008) Environmental awareness, interests and motives of botanic garden visitors: implications for interpretive practice. *Tourism Management* 29(3), 439–444.

Banks, S.B. (2015) Managing risks from hazardous substances in the Economic Botany Collection at the Royal Botanic Gardens, Kew: a pragmatic approach. *Journal of the Institute of Conservation* 38(2), 130–145.

Barley, R. (2019) Oral Presentation on behalf of Kew Gardens to the UK Parliamentary Sub-Committee on Digital Media, Culture and Sport Committee on Garden Design and Tourism. Fourteenth Report of Session 2017–19. Available at: https://publications.parliament.uk/pa/cm201719/cmselect/cmcumeds/2002/2002.pdf (accessed August 1, 2020).

Barton, E.T. (2017) Engaging millennials in public gardens through digital and social media strategies. MSc. thesis, University of Delaware, Newark, Delaware.

Bartram's Garden (Philadelphia, Pennsylvania). https://bartramsgarden.org/ (accessed August 10, 2020).

Batra, R., Lehmann, D.R. and Singh, D. (1993) The brand personality component of brand goodwill: some antecedents and consequences. In: Aaker, D.A. and Biel, A. (eds) *Brand Equity and Advertising*. Lawrence Erlbaum Associates, Hillsdale, New Jersey, pp. 83–96.

Bellamy-Ferriday House & Garden (Bethlehem, Connecticut). https://ctlandmarks.org/properties/bellamy-ferriday-house-garden/ (accessed August 10, 2020).

Beyer, K.M.M., Kaltenbach, A., Szabo, A., Bogar, S., Nieto, F.J. and Malecki, K.M. (2014) Exposure to neighborhood green space and mental health: evidence from the Survey of the Health of Wisconsin. *International Journal of Environmental Research and Public Health* 11(3), 3453–3472.

Benfield, R.W. (2001) 'Good things come to those who wait': sustainable tourism and timed entry at Sissinghurst Castle Garden, Kent. *Tourism Geographies* 3(2), 207–217.

Benfield, R.W. (2013) *Garden Tourism*. CAB International, Wallingford, UK.

Benfield, R. (2014) Economic Impact Assessment of the Arid Lands Botanic Garden, Port Augusta, South Australia. Unpublished report for the Economic Development Board of the City of Port Augusta, South Australia.

Berthon, P., Hulbert, J.M. and Pitt, L.F. (1999) Brand Management Prognostications. *Sloan Management Review* 40(2), 53–65.

Bethune, A. (2018) 4 important place branding questions. Available at: https://www.originoutside.com/insights/4-important-place-branding-questions (accessed July 20, 2020).

Blain, C., Levy, S.E. and Brent Ritchie, J.R. (2005) Destination branding: insights and practices from destination management organizations. *Journal of Travel Research* 43(4), 328–338.

Bogor Botanic Gardens (Bogor, Indonesia). https://whc.unesco.org/en/tentativelists/6353/ (accessed August 10, 2020).

Boorstin, D.J. (1964) *The Image: A Guide to Pseudo-events in America*. Harper & Row, New York.

Botanical Garden of Rio de Janeiro (Jardim Botânico do Rio de Janeiro, Brazil). http://en.jbrj.gov.br/ (accessed August 10, 2020).

Botanic Gardens Conservation International (BGCI). Available at: https://www.bgci.org (accessed July 20, 2020).

Brooklyn Botanic Garden (Brooklyn, New York). Available at: https://www.bbg.org (accessed 20 July 2020).

Blue Sail Cheshire Gardens of Distinction (2012) 2011 Impact evaluation. Unpublished report for Visit Chester and Cheshire.

Brand New (2015) The Royal Treatment – New Logo and Identity for Royal Botanic Gardens, Kew by Pentagram. Available at: https://www.underconsideration.com/brandnew/archives/new_logo_and_identity_for_royal_botanic_gardens_kew_by_pentagram.php (accessed July 30, 2020).

Briesch, R.A. (2017) Dallas Arboretum Economic Impact Study. Unpublished report, Southern Methodist University, Dallas, Texas.

The Butchart Gardens (Victoria, British Columbia, Canada). https://www.butchartgardens.com/ (accessed August 10, 2020).

Buck, D. (2016) Gardens and Health: Implications for Policy and Practice. Report by the King's Fund produced for the National Garden Scheme. Available at: https://ngs.org.uk/wp-content/uploads/2019/06/Kings-Fund-Report-1.pdf (accessed July 31, 2020).

The Burra Charter (2013) The Burra Charter: The Australia ICOMOS Charter for Places of Cultural Significance, 2013. Available at: https://australia.icomos.org/wp-content/uploads/The-Burra-Charter-2013-Adopted-31.10.2013.pdf (accessed July 27, 2020).

Byun, J. and Jang, S. (2015) Effective promotions for membership subscriptions and renewals to tourist attractions: discount vs. bonus. *Tourism Management* 50, 194–203.

Cactus Garden (Jardín de Cactus) (Guatiza, Lanzarote, Spain). https://www.cactlanzarote.com/en/cact/jardin-de-cactus/ (accessed August 10, 2020).

Cameron, R. and Griffiths, A. (2016) Gardening; its value in terms of human health and well-being. Presented at Conference on Health and Happiness, Hampton Court Palace, London, July 2016.

Cameron, R.W.F. and Hitchmough, J.D. (2016) *Environmental Horticulture: Science and Management of Green Landscapes*. CAB International, Wallingford, UK.

Cengage Learning (undated) Case Study 4.1W: Motivations for visiting gardens in Britain. In Chapter 4: Understanding the tourist as a consumer. Available at: http://cws.cengage.co.uk/page2/students/cases/4-1.pdf (accessed August 1, 2020).

Chandler, V (2017) The six biggest gin drinking trends of 2018. *Good Housekeeping*, August 18. Available at: https://www.goodhousekeeping.com/uk/food/a570805/gin-trends/ (accessed July 26, 2020).

Changi Airport gardens (Singapore). https://www.changiairport.com/en/discover/recommended-itineraries/go-on-a-nature-trail.html (accessed August 10, 2020).

Chester Zoo and botanical gardens (Chester, UK). https://www.chesterzoo.org/our-zoo/plants-and-gardens/ (accessed August 10, 2020).

Chicago Botanic Garden (Glencoe, Illinois). Available at: https://www.chicagobotanic.org/ (accessed July 20, 2020).

Chmura Economics & Analytics (2016) *Economic Impact of Virginia Historic Garden Week*. Survey for the Garden Clubs of Virginia. Chmura Economics & Analytics, Richmond, Virginia.

Cleveland Botanical Garden (Cleveland, Ohio). Available at: http://www.cbgarden.org (accessed July 20, 2020).

Connell, J. (2002) A critical analysis of gardens as a tourism and recreation resource in the UK. PhD thesis, University of Plymouth, Plymouth, UK.

Connell, J. (2004) The purest of human pleasures: the characteristics and motivations of garden visitors in Great Britain. *Tourism Management* 25(2), 229–247.

Connell, J., Page, S.J. and Meyer, D. (2015) Visitor attractions and events: responding to seasonality. *Tourism Management* 46, 283–298.

Cosgrove, S. (2017) How to build a garden brand. *Horticulture Week*. Available at: https://www.hortweek.com/build-garden-brand/parks-and-gardens/article/1432897 (accessed July 20, 2020).

Cotswold Country Gardens at Barnsley House (Gloucestershire, UK). https://www.barnsleyhouse.com/gardens/ (accessed August 19, 2020).

Croucher, K.L., Myers, L. and Bretherton, J. (2008) *The Links Between Green Space and Health: A Critical Literature Review*. Greenspace Scotland, Stirling, UK.

Culture Track Research (2020) Culture and community in a time of crisis. Culture Track. Available at: http://www.culturetrack.com/research/covidstudy/ (accessed October 20, 2020).

Cunningham, S. and Charlier, J. (2019) *Buffalo-Style Gardens: Create a Quirky, One-of-a-Kind Private Garden with Eye-Catching Designs*. St. Lynn's Press, Pittsburgh, Pennsylvania.

Curtin, S. and Fox, D. (2014) Human dimensions of wildlife gardening: its development, controversies and psychological benefits. In: Dixon, G.R. and Aldous, D.E. (eds) *Horticulture: Plants for People and Places*, Volume 3. Springer, Heidelberg, Germany, pp. 1025–1047.

Dallas Arboretum and Botanical Garden (Dallas, Texas). Available at: https://www.dallasarboretum.org/ (accessed July 30, 2020).

Daniel Stowe Botanical Garden (Belmont, North Carolina). https://www.dsbg.org/ (accessed August 10, 2020).

Dasgupta, P. and Serageldin, I. (eds) (1999) *Social Capital: A Multifaceted Perspective*. World Bank Group, Washington, DC. Available at: http://documents.worldbank.org/curated/en/663341468174869302/Social-capital-a-multifaceted-perspective (accessed August 1, 2020).

David Austin Rose Gardens (Albrighton, UK). https://www.davidaustinroses.co.uk/pages/david-austin-rose-gardens (accessed August 10, 2020).

Delaware Botanic Gardens at Pepper Creek (Dagsboro, Delaware). http://www.delawaregardens.org/ (accessed August 10, 2020).

Demby, E.W. (1994) Psychographics revisited: the rebirth of a technique. *Marketing Research* 26(2), 26–29.

Denver Botanic Gardens (Denver, Colorado). Available at: https://www.botanicgardens.org/ (accessed July 21, 2020).

Department of Parks and Recreation (2020) *Visitor Numbers, Botanical Gardens*. City and County of Honolulu, Hawaii.

Depledge, M.H., Stone, R.J. and Bird. W.J. (2011) Can natural and virtual environments be used to promote improved human health and wellbeing? *Environmental Science & Technology* 45(11), 4660–4665.

Desert Botanical Garden (Phoenix, Arizona). Available at: https://www.dbg.org/ (accessed July 21, 2020).

De Vere, N., Jones, L., Gilmore, T., Moscrop, J., Lowe, A., *et al.* (2017) Using DNA metabarcoding to investigate honey bee foraging reveals limited flower use despite high floral availability. *Scientific Reports* 7, 42838. DOI: 10.1038/srep42838

Dodd, J. and Jones, C. (2010) Towards a New Social Purpose – Redefining the Role of Botanic Gardens. Report of the Calouste Gulbenkian Foundation for Botanic Gardens Conservation International, Kew, London. Available at: https://www.bgci.org/resources/bgci-tools-and-resources/towards-a-new-social-purpose-redefining-the-role-of-botanic-gardens/ (accessed July 30, 2020).

Dodd, J. and Jones, C. (2011) Towards a new social purpose: the role of botanic gardens in the 21st century. *Roots* 8(1). Available at: https://www.bgci.org/wp/wp-content/uploads/2019/04/Roots_8.1.pdf (accessed July 21, 2020).

Drea, S. (2011) The end of the Botany degree in the UK. *Bioscience Education* 17(1), 1–7. https://doi.org/10.3108/beej.17.2

Drucker, P.F. (1954) *The Practice of Management*. Harper & Row, New York.

Dubai Miracle Garden (Dubai, United Arab Emirates). https://www.dubaimiraclegarden.com/ (accessed August 10, 2020).

Eden Project (Bodelva, UK). https://www.edenproject.com/ (accessed August 10, 2020).

Emmett, R.S. (2016) *Cultivating Environmental Justice: A Literary History of US Garden Writing*. University of Massachusetts Press, Amhurst, Massachusetts.

Fáilte Ireland (2016) Cultural product usage among overseas tourists in 2014. Available at: https://www.failteireland.ie/FailteIreland/media/WebsiteStructure/Documents/3_Research_Insights/1_Sectoral_SurveysReports/Cultural-activity-product-usage-among-overseas-tourists-in-2014.pdf?ext=.pdf (accessed July 21, 2020).

Fairchild Tropical Botanic Garden (Coral Gables, Florida). Available at: https://www.fairchildgarden.org/ (accessed July 20, 2020).

Farm & Gardens at Thomas Jefferson Monticello (Charlottesville, Virginia). https://www.monticello.org/house-gardens/farms-gardens/ (accessed August 10, 2020).

First Nations Garden (Montreal, Quebec, Canada). https://espacepourlavie.ca/en/first-nations-garden#:~:-text=The%20First%20Nations%20Garden%2C%20open,the%20First%20Nations%20of%20Qu%C3%A9bec. (accessed August 10, 2020).

Fiveash, R. (2018) The branding of botanical gardens for the 21st century. Master's thesis, Central Connecticut State University, New Britain, Connecticut.

Fox, D. (2007) Understanding garden visitors: the affordances of a leisure environment. PhD thesis, University of Bournemouth, Bournemouth, UK.

Fox, D. (2015) Making a difference by understanding what makes people visit gardens. Presented at the North American Garden Tourism Conference, Toronto, Canada, 16–18 March 2015.

Fox, D. (2016) Leisure time preference: the influence of gardening on garden visitation. Presented at the LARASA World Leisure Congress, Durban, South Africa, 27–29 June 2016.

Fox, D. (2017a) The current state of academic research in garden tourism. Presented at International Garden Tourism Conference, Kew Gardens, London, May 19.

Fox, D. (2017b) External agents of change: a ten-year trend study of USA garden visitor behaviour in England. *Tourism Recreation Research* 42(4), 446–456.

Fox, D. and Edwards, J.R. (2008) Managing gardens. In: Fyall, A., Leask, B., Garrod, S. and Wanhill, S. (eds) *Managing Visitor Attractions*, 2nd edn. Butterworth-Heinemann, Oxford, pp. 217–236.

Fox, D. and Edwards, J.R. (2009) A preliminary analysis of the market for small, medium and large horticultural shows in England. *Event Management* 12(3/4), 199–208.

Fox, D., Johnson, N. and Wallace, S. (2009) Designing the past: the National Trust as social-material agency. In: Glynne, J., Hackney, F. and Minton, V. (eds) *Networks of Design: Proceedings of the 2008 Annual International Conference of the Design History Society*. BrownWalker Press, Boca Raton, Florida, pp. 164–168.

Fox, D., Edwards, J. and Wilkes, K. (2010) Employing the 'grand tour' approach to aid understanding of garden visiting. In: Richards, G. and Munsters, W. (eds) *Cultural Tourism Research Methods*. CAB International, Wallingford, UK, pp. 75–86.

Franklin Park Conservatory and Botanical Gardens (Columbus, Ohio). Available at: https://www.fpconservatory.org/ (accessed July 21, 2020).

Frey, W. (2011) *Melting Pot Cities and Suburbs: Racial and Ethnic Change in Metro America in the 2000s*. Brookings Institute, Washington, DC.

The Garden at Chatsworth (Derbyshire, UK). https://www.chatsworth.org/garden/ (accessed August 10, 2020).

The Garden at Newfields (Indianapolis, Indiana). Available at: https://discovernewfields.org/do-and-see/places-to-go/garden (accessed July 21, 2020).

Garden Museum (London). https://gardenmuseum.org.uk/ (accessed August 10, 2020).

Garden of Cosmic Speculation (Dumfries, UK). https://gardenofcosmicspeculation.com/ (accessed August 10, 2020).

The Gardens at George Washington's Mount Vernon (Mount Vernon, Virginia). https://www.mountvernon.org/the-estate-gardens/gardens-landscapes/ (accessed August 10, 2020).

Gardens Buffalo Niagara (Buffalo, New York). https://www.gardensbuffaloniagara.com/ (accessed August 10, 2020).

Gardens by the Bay (Singapore). https://www.gardensbythebay.com.sg/ (accessed August 10, 2020).

Gardens on Spring Creek (Fort Collins, Colorado). https://www.fcgov.com/gardens/ (accessed August 10, 2020).

Garfield Park Conservatory (Chicago, Illinois). https://garfieldconservatory.org/ (accessed August 10, 2020).

Getz, D. and Page, S.J. (2016a) *Event Studies: Theory, Research and Policy for Planned Events*, 3rd edn. Routledge, London.

Getz, D. and Page, S.J. (2016b) Progress and prospects for event tourism research. *Tourism Management* 52 (February), 593–631.

Glenstone (Potomac, Maryland). Available at: https://www.glenstone.org/ (accessed August 5, 2020).

Goldsmith, E. (1972) *A Blueprint for Survival*. Tom Stacey Ltd, London.

Goody, J. (1993) *The Culture of Flowers*. Cambridge University Press, Cambridge.

Government of Ontario, Canada, Ministry of Tourism (2007) 2006 TAMS Survey; Travel and Motivation Survey of Americans. Available at: http://www.mtc.gov.on.ca/en/research/travel_activities/TAMS%20 2006%20Overview%20U.S.%20Report%20(FINAL).pdf (accessed July 30, 2020).

Gravetye Manor Gardens (East Grinstead, UK). https://www.gravetyemanor.co.uk/the-gardens/ (accessed August 10, 2020).

Great British Gardens. Available at: https://www.greatbritishgardens.co.uk/wildlife-gardens.html (accessed July 26, 2020).

Greater Des Moines Botanical Garden (Des Moines, Iowa). https://www.dmbotanicalgarden.com/ (accessed August 10, 2020).

Hall Kelly, M. (2017) *Lilac Girls: A Novel*. Penguin Random House, New York.

Hall Kelly, M. (2020) *Lost Roses: A Novel*. Penguin Random House, New York.

Halverson Group (2016) Creating a Picture for the Future of the IMA. Jobs-to-be-Won™ Segmentation Results. Unpublished study for the Indianapolis Museum of Art, Indianapolis, Indiana.

Hamilton Gardens (Auckland, New Zealand). https://hamiltongardens.co.nz/ (accessed August 10, 2020).

Hankinson, G. (2007) The management of destination brands: five guiding principles based on recent developments in corporate branding theory. *Journal of Brand Management* 14(3), 240–254.

Hartig, T., Mitchell, R., de Vries, S. and Frumkin, H. (2014) Nature and health. *Annual Review of Public Health* 35, 207–228.

Hatam National Botanical Garden (Pretoria, South Africa). https://www.sanbi.org/gardens/hantam/ (accessed August 10, 2020).

Hatch, M.J. and Schultz, M. (2003) Bringing the corporation into corporate branding. *European Journal of Marketing* 37(7/8), 1041–1064.

He, H. and Chen, J. (2012) Educational and enjoyment benefits of visitor education centers at botanical gardens. *Biological Conservation* 149(1), 103–112.

Heywood, V.H. (2017) The future of plant conservation and the role of botanic gardens. *Plant Diversity* 39(6), 309–313.

Hollister House Garden (Washington, Connecticut). https://hollisterhousegarden.org/ (accessed August 10, 2020).

Ho'omaluhia Botanical Garden (Kaneohe, Hawaii). https://www.honolulu.gov/parks/hbg.html?id=569:ho (accessed August 10, 2020).

Horwath HTL (2008) Economic Assessment of Hamilton Botanic Gardens. Unpublished report for Waikato City Council, New Zealand.

Hosany, S., Ekinci, Y. and Uysal, M. (2007) Destination image and destination personality. *International Journal of Culture, Tourism and Hospitality Research* 1(1), 62–81.

Howarth, C. (undated) Brands we love – royal botanic gardens. *Truly Deeply Blog*. Available at: http://www.trulydeeply.com.au/2017/06/brands-we-love-royal-botanic-gardens/ (accessed July 21, 2020).

Hu, M.X. (2017) Botanical gardens – build a Noah arc for plants (in Chinese). Available at: http://tech.sina.com.cn/d/2017-11-17/doc-ifynwxum2160850.shtml (accessed February 20, 2019).

Hulme, P.E. (2015) Resolving whether botanic gardens are on the road to conservation or a pathway for plant invasions. *Conservation Biology* 29(3), 816–824. DOI: 10.1111/cobi.12426

Idaho Botanical Garden (Boise, Idaho). https://idahobotanicalgarden.org/ (accessed August 10, 2020).

Inala Jurassic Garden (Bruny Island, Tasmania, Australia). https://www.inalanaturetours.com.au/bruny-island/inala-jurassic-garden (accessed August 10, 2020).

Inverewe Garden (Poolewe, UK). https://www.nts.org.uk/visit/places/inverewe (accessed August 10, 2020).

Ivarsson, C.T and Hagerhall, C.M. (2008) The perceived restorativeness of gardens – assessing the restorativeness of a mixed built and natural scene type. *Urban Forestry & Urban Greening* 7, 107–118.

Jacob, S. and McClintock, M.K. (2000) Psychological state and mood effects of steroidal chemosignals in women and men. *Hormones and Behavior* 37(1), 57–78.

Jardin Botánico Canario Viera y Clavijo (Las Palmas de Gran Canaria, Spain). http://www.jardincanario.org/en (accessed August 11, 2020).

Jardin Botanico – Dr. Alfredo Barrera Marin (Chetumal, Mexico). https://tools.bgci.org/garden.php?id=1218 (accessed August 10, 2020).

Keller, K.L. (1993) Conceptualizing, measuring and managing customer-based brand equity. *Journal of Marketing* 57(1), 1–22.

Kenilworth Castle and Elizabethan Garden (Warwickshire, UK). https://www.english-heritage.org.uk/visit/places/kenilworth-castle/ (accessed August 10, 2020).

Kirstenbosch National Botanical Garden (Cape Town, South Africa). https://www.sanbi.org/gardens/kirstenbosch/ (accessed August 10, 2020).

Kohlleppel, T., Bradley, J. and Jacob, S.A. (2001) Walk through the garden: can a visit to a botanic garden reduce stress? *HortTechnology* 12(3), 489–492.

Krinsky, J. and Simonet, M. (2011) Safeguarding private value in public spaces: the neoliberalization of public service work in New York City's parks. *Social Justice* 38(1/2), 28–47.

Kunming Botanical Garden (Kunming, China). https://tools.bgci.org/garden.php?id=337 (accessed August 10, 2020).

Kwelera National Botanical Garden (East London, South Africa). https://www.sanbi.org/gardens/kwelera/ (accessed August 10, 2020).

Latour, B. (1991) *We Have Never Been Modern*, tr. Catherine Porter (1993). Harvester, Brighton, UK.

Lauritzen Gardens (Omaha, Nebraska). https://www.lauritzengardens.org/ (accessed August 10, 2020).

Lee, G., Tussyadiah, I.P. and Zach, F. (2010) A visitor-focused assessment of new product launch: the case of Quilt Gardens Tour[SM] in Northern Indiana's Amish country. *Journal of Travel & Tourism Marketing* 27, 723–735.

Lewis Ginter Botanical Garden (Richmond, Virginia). Available at: https://www.lewisginter.org (accessed July 21, 2020).

Lin, Y.-H., Tsai, C.-C., Sullivan, W.C., Chang, P.-J. and Chang, C.-Y. (2014) Does awareness affect the restorative function and perception of street trees? *Frontiers in Psychology* 5, 906.

Lipovská, B. (2013) The fruit of garden tourism may fall over the wall: small private gardens and tourism. *Tourism Management Perspectives* 6, 114–121.

Lippincott Corporation (2015) 20 milestones in the history of branding. Exhibition at the Design Museum, London, September 19–27, 2015. *Creative Bloq (Computer Arts)*, September 18, 2015. Available at: https://www.creativebloq.com/branding/milestones-history-branding-91516855 (accessed August 5, 2020).

Longwood Gardens (Kennett Square, Pennsylvania). https://longwoodgardens.org/ (accessed August 10, 2020).

Loughran, K. (2014) Parks for profit: the High Line, growth machines, and the uneven development of open spaces. *City and Community* 13(1), 49–68.

Lyndhurst Mansion (Tarrytown, New York). https://lyndhurst.org/ (accessed August 10, 2020).

MacCannell, D. (1976) *The Tourist; A New Theory of the Leisure Class*. University of California Press, Berkley, California.

McAuley, T.E. (2016) Viewing a myriad leaves: Man'yō Botanical Gardens in Japan. *International Journal of Contents Tourism* 1(2), 1–16.

McClintock, M. (1971) Menstrual synchrony and suppression. *Nature* 229, 244–245.

McIntire Botanical Garden (Charlottesville, Virginia). Available at: https://www.mcintirebotanicalgarden.org (accessed July 21, 2020).

Malone-France, K. and Thompson, M. (2014) When historic buildings and landscapes ARE the museum collection. *History News* 21, 21–27. Available at: http://download.aaslh.org/history+news/2014_Summer_When+Historic+Buildings+and+Landscapes+are+the+Museum+Collection_Malone-France.pdf (accessed July 21, 2020).

Management Study Guide. Available at: www.managementstudyguide.com (accessed July 21, 2020).

Markwell, K. (2014) Book review. *Annals of Leisure Research* 17(4), 499–500.

Mehaffey, S. (undated) Higher Unity: A Manifesto for Stewarding and Activating the Landscape of Farnsworth House. Unpublished paper for the National Trust for Historic Preservation, Washington, DC.

Millington, N. (2015) From urban scar to "park in the sky": terrain vague, urban design, and the remaking of New York City's High Line Park. *Environment and Planning A* 47(11), 2324–2338.

Missouri Botanical Garden (St. Louis, Missouri). Available at: http://www.missouribotanicalgarden.org (accessed July 21, 2020).

Mitchell, R. and Popham, F. (2008) Effect of exposure to natural environment on health inequalities; an observational population study. *Lancet* 372(9650), 1655–1660.

Monroe, B. (2018) Behind the scenes: a brand blooms at Lewis Ginter Botanical Garden. *Public Garden Magazine* 33(3).

Montreal Botanical Garden (Jardin botanique de Montréal) (Montreal, Quebec, Canada). Available at: https://m.espacepourlavie.ca/en/botanical-garden (accessed July 21, 2020).

Morton, J. (2014) What are the five senses of flowers? Available at: http://www.ehow.com/list_7594946_five-senses-flowers.html (accessed May 26, 2014).

Moskwa, E.C. and Crilley, G. (2012) Recreation, education, conservation: the multiple roles of botanic gardens in Australia. *Annals of Leisure Research* 15(4), 404–421. DOI: 10.1080/11745398.2012.744276

Mt. Cuba Center (Hockessin, Delaware). https://mtcubacenter.org/ (accessed August 10, 2020).

Museums for All. https://museums4all.org/ (accessed July 24, 2020).

Nadel-Klein, J. (2010a) Cultivating taste and class in the garden. In: Grønseth, A.S. and Davis, D.L. (eds) *Mutuality and Empathy: Self and Other in the Ethnographic Encounter*. Sean Kingston, Canon Pyon, UK, pp. 107–121.

Nadel-Klein, J. (2010b) Gardening in time: happiness and memory in American horticulture. In: Collins, P. and Gallinat, A. (eds) *The Ethnographic Self as Resource: Writing Memory and Experience into Ethnography*. Berghahn, New York and Oxford, pp. 165–184.

National Botanic Garden of Wales (Carmarthenshire, UK). https://botanicgarden.wales/ (accessed August 10, 2020).

National Tropical Botanical Garden of Hawaii (Hawaii). https://ntbg.org/ (accessed August 10, 2020).

National Wildlife Federation. https://www.nwf.org/search?col=12&col=3&col=7&query=Birds&advanced=false (accessed July 26, 2020).

New York Botanical Garden (NYBG) (New York). Available at: https://www.nybg.org (accessed July 21, 2020).

Ningbo Botanical Garden (Ningbo, China). https://tools.bgci.org/garden.php?id=5314&ftrCountry=CN&ftrKeyword=&ftrBGCImem=&ftrIAReg= (accessed August 10, 2020).

Nong Nooch Tropical Garden (Chon Buri, Thailand). Available at: http://www.nongnoochtropicalgarden.com/ (accessed August 10, 2020).

Nordh, H., Hartig, T., Hagerhall, C.M. and Fry, G. (2009) Components of small urban parks that predict the possibility for restoration. *Urban Forestry & Urban Greening* 8(4), 225–235.

Norfolk Botanical Garden (Norfolk, Virginia). https://norfolkbotanicalgarden.org/ (accessed August 10, 2020).

Ogunseitan, O.A. (2005) Topophilia and the quality of life. *Environmental Health Perspectives* 113(2), 143–148. DOI: 10.1289/ehp.7467.

Oman Botanic Garden (Muscat, Oman). https://omanbotanicgarden.om/ (accessed August 10, 2020).

Oxford Economics (2018) *The Economic Impact of Ornamental Horticulture and Landscaping in the UK*. A report for the for the Ornamental Horticulture Roundtable Group. Oxford Economics, London.

Pérez-Sanagustín, M., Parra, D., Verdugo, R. and Garcia-Galleguillos, G. (2016) Using QR codes to increase user engagement in museum-like spaces. *Computers in Human Behaviour* 60, 73–85.

Perkins, A., Hamnett, S., Pullen, S., Zito, R. and Trebilcock, D. (2009) Transport, housing and urban form: the life cycle energy consumption and emissions of city centre apartments compared with suburban dwellings. *Urban Policy and Research* 27(4), 377–396. https://doi.org/10.1080/08111140903308859

Pew (2019) Defining generations: where Millennials end and Generation Z begins. Available at: https://www.pewresearch.org/fact-tank/2019/01/17/where-millennials-end-and-generation-z-begins/ (accessed August 1, 2020).

Phipps Conservatory and Botanical Gardens (Pittsburgh, Pennsylvania). https://www.phipps.conservatory.org/ (accessed August 10, 2020).

Pine, J.B. and Gilmore, J.H. (1999) *The Experience Economy*. Harvard Business School Press, Boston, Massachusetts.

Pitmedden Garden (Pitmedden, UK). https://www.nts.org.uk/visit/places/pitmedden-garden (accessed August 10, 2020).

Pittsburgh Botanic Garden (Oakdale, Pennsylvania). https://pittsburghbotanicgarden.org/ (accessed August 10, 2020).

Plog, S.C. (1974) Why destination areas arise and fall in popularity. *Cornell Hotel and Restaurant Administration Quarterly* 14(4), 55–58.

Plog, S.C. (1991) *Leisure Travel: Making It a Growth Market ... Again!* Wiley, New York.

Pollan, M. (1991) *Second Nature; A Gardener's Education*. Dell, New York.

Powell Gardens, Kansas City's Botanical Garden (Kingsville, Missouri). https://powellgardens.org/ (accessed August 10, 2020).

Powney, G.D., Carvell, C., Edwards, M., Morris, R.K.A., Roy, H.E., *et al.* (2019) Widespread losses of pollinating insects in Britain. *Nature Communications* 10, 1018. DOI: 10.1038/s41467-019-08974-9

Pritchard, A. and Morgan, N. (2010) 'Wild on' the beach: discourses of desire, sexuality and liminality. In: Waterton, E. and Watson, S. (eds) *Culture, Heritage and Representation: Perspectives on Visuality and the Past*. Routledge, Abingdon, UK, pp. 127–143.

Qinling National Botanical Garden (Xi'an, China). https://tools.bgci.org/garden.php?id=4554 (accessed August 10, 2020).

Queens Botanical Garden (Flushing, New York). https://queensbotanical.org/ (accessed August 10, 2020).

Quito Botanical Garden (Jardín Botánico de Quito). http://www.jardinbotanicoquito.com/es/ (accessed August 10, 2020).

Reiman Gardens (Ames, Iowa). https://www.reimangardens.com/ (accessed August 10, 2020).

Reiner, T.A. and Wilson, R.H. (1979) Planning and decision-making in the Soviet city: rent, land and urban form. In: French, R.A. and Hamilton, R.E.I. (eds) *The Socialist City: Spatial Structure and Urban Policy*. Wiley, London, pp. 49–71.

Ries, A. and Ries, L. (1998) *22 Immutable Laws of Branding: How to Build a Product or Service into a World-Class Brand*. HarperCollins, New York.

RHS Wisley (Wisley, UK). https://www.rhs.org.uk/gardens/wisley (accessed August 10, 2020).

Rockport Analytics (2014) The Economic & Fiscal Impacts of the Delaware Botanic Garden at Pepper Creek. Available at: https://static1.squarespace.com/static/55898a17e4b032af2c554890/t/55c62 660e4b0851a71d150c3/1439049312122/Economic+Impact+of+Delaware+Botanic+Garden+Updated+ FINAL+8-5-2015.pdf (accessed August 1, 2020).

Rosenblum, L.D. (2010) *The Extraordinary Powers of our Five Senses: See What I'm Saying*. Norton, New York.

Royal Botanical Gardens (Hamilton, Ontario, Canada). https://www.rbg.ca/ (accessed August 10, 2020).

Royal Botanic Garden Edinburgh (RGBE) (Edinburgh). https://www.rbge.org.uk (accessed August 1, 2020).

Royal Botanic Garden Sydney (Sydney, New South Wales, Australia). https://www.rbgsyd.nsw.gov.au/ (accessed August 10, 2020).

Royal Botanic Gardens, Kew (London). Available at: https://www.kew.org/ (accessed August 1, 2020).

Royal Horticultural Society (RHS). https://www.rhs.org.uk/ (accessed August 10, 2020).

San Antonio Botanical Garden (San Antonio, Texas). https://www.sabot.org/ (accessed August 10, 2020).

San Francisco Botanical Garden (San Francisco, California). Available at: https://www.sfbg.org (accessed August 1, 2020).

Santa Fe Botanical Garden (Santa Fe, New Mexico). https://santafebotanicalgarden.org/ (accessed August 10, 2020).

Sayorwan, W., Siripornpanich, V., Piriyapunyaporn, T., Hongratanaworakit, T., Kotchabhakdi, N. and Ruangrungsi, N. (2012) The effects of lavender oil inhalation on emotional states, autonomic nervous system, and brain electrical activity. *Journal of the Medical Association of Thailand* 95(4), 598–606.

Seligman, M. (2002) *Authentic Happiness: Using New Positive Psychology to Realize Your Potential for Lasting Fulfillment.* Free Press, New York.

Senteurs d'Angkor (Siem Riep, Cambodia). https://senteursdangkor.com/ (accessed August 19, 2020).

Shanghai Chenshan Botanical Garden (Shanghai, China). http://en.csnbgsh.cn/sites/chenshan/chenshan_en/index.ashx (accessed August 10, 2020).

Singapore Botanic Gardens (Singapore). Available at: https://www.nparks.gov.sg/sbg (accessed July 21, 2020).

Sir Seewoosagur Ramgoolam Botanical Garden (Port Louis, Mauritius). http://ssrbg.govmu.org/English/garden/Pages/default.aspx (accessed August 10, 2020).

Smith, A. (2019) Event takeover? The commercialization of London's parks. In: Smith, A. and Graham, A. (eds) *Destination London: The Expansion of the Visitor Economy.* University of Westminster Press, London, pp. 205–223. DOI: 10.16997/book35.j

Smith, M.K. (2000–2009) Social capital. *The Encyclopedia of Pedagogy and Informal Education.* Available at: https://infed.org/mobi/social-capital/ (accessed August 12, 2020).

Smith, T. (2018) Lewis Ginter Botanical Garden updates its brand image. *Richmond Times-Dispatch.* Available at: https://richmond.com/business/lewis-ginter-botanical-garden-updates-its-brand-image/article_497cb9c9-3b13-513a-9afc-d005e6b065f5.html (accessed July 22, 2020).

Sri Lanka Department of National Botanical Gardens. http://msdw.gov.lk/departments/department-of-national-botanical-gardens/ (accessed August 10, 2020).

State Arboretum of Virginia (Winchester, Virginia). http://blandy.virginia.edu/arboretum (accessed August 10, 2020).

State Botanical Garden of Georgia (Athens, Georgia). https://botgarden.uga.edu/ (accessed August 10, 2020).

Stourhead (Wiltshire, UK). https://www.nationaltrust.org.uk/stourhead (accessed August 10, 2020).

Sweet, W. (n.d.) Jeremy Bentham 1748–1832. *Internet Encyclopedia of Philosophy.* Available at: https://www.iep.utm.edu/bentham/ (accessed August 1, 2020).

Terra Nostra Garden (Furnas, Azores, Portugal). http://www.parqueterranostra.com/en-us/botanicallibrary.aspx (accessed August 10, 2020).

Thalassinou, M. (2018) Wind of change rustles leaves at Sydney's Royal Botanic Garden. *Transform Magazine*, July 23. Available at: https://www.transformmagazine.net/articles/2018/wind-of-change-rustles-leaves-at-sydney-s-royal-botanic-garden/ (accessed August 1, 2020).

Toronto Botanical Garden (Toronto, Ontario, Canada). https://torontobotanicalgarden.ca/ (accessed August 10, 2020).

Tourism New Zealand (2015) *International Visitor Survey 2013.* Ministry of Business, Innovation and Employment, Auckland, New Zealand.

Tower Hill Botanic Garden (Worcester, Massachusetts). https://www.towerhillbg.org/ (accessed August 10, 2020).

Travis, D. (2000) *Emotional Branding: How Successful Brands Gain the Irrational Edge.* Prima Publishing, Roseville, California.

Tuan, Y.F. (1974) *Topophilia: A Study of Environmental Perception, Attitudes, and Values.* Prentice-Hall, Englewood Cliffs, New Jersey.

Turner, L. and Ash, J. (1975) *The Golden Hordes: International Tourism and the Pleasure Periphery.* Constable, London.

Twinings. Available at: https://www.twinings.co.uk/ (accessed August 1, 2020).

UK Heritage Lottery Fund (2016) State of UK Public Parks 2016. Available at: https://www.heritagefund.org.uk/publications/state-uk-public-parks-2016 (accessed August 13, 2020).

UK Parliament (2019) Report on Garden Design and Tourism by the Sub-Committee on Digital Media, Culture and Sport. Fourteenth Report of Session 2017–19. Available: https://publications.parliament.uk/pa/cm201719/cmselect/cmcumeds/2002/2002.pdf (accessed August 1, 2020).

Ulrich, R.S. (1983) Aesthetic and affective response to natural environment. In: Altman, I. and Wohlwill, J.F. (eds) *Behavior and the Natural Environment. Human Behavior and Environment*, Volume 6. Plenum Press, New York, pp. 85–125.

UNAM Botanical Garden (Mexico City). https://www.atlasobscura.com/places/unam-botanical-garden (accessed August 10, 2020).

United Nations World Tourism Organization (UNWTO). Available at: http://www2.unwto.org/ (accessed July 21, 2020).

United States Botanic Garden (Washington, DC). Available at: https://www.usbg.gov/ (accessed July 21, 2020).

United States Census Bureau (2012) Statistical Abstract of the United States: 2012. Section 26. Arts, Recreation, and Travel. Available at: https://www.census.gov/library/publications/time-series/statistical_abstracts.html https://www.census.gov/library/publications/2011/compendia/statab/131ed/arts-recreation-travel.html (accessed August 12, 2020).

United States National Arboretum (Washington, DC). https://www.usna.usda.gov/ (accessed August 10, 2020).

University of Alberta Botanic Garden (Edmonton, Alberta, Canada). https://botanicgarden.ualberta.ca/ (accessed August 10, 2020).

University of California, Riverside (UCR) Botanic Gardens (Riverside, California). https://gardens.ucr.edu/ (accessed August 10, 2020).

Urry, J. (1990) *The Tourist Gaze: Leisure and Travel in Contemporary Societies*. Sage, London.

Urry, J. and Larsen, J. (2011) *The Tourist Gaze 3.0*, 3rd ed. Sage, London.

Vallarta Botanical Gardens (Puerto Vallarta, Mexico). https://www.vbgardens.org/ (accessed August 10, 2020).

Valley of Flowers in Nanda Devi National Park (India). https://whc.unesco.org/en/list/335/ (accessed August 10, 2020).

VisitBritain (2015a) Foresight report on activities undertaken in Britain 2015. Available at: https://www.visitbritain.org/activities-undertaken-britain (accessed June 18, 2015).

VisitBritain (2015b) What are potential visitors to Britain interested in doing? *Foresight* – issue 135. Available at: https://www.visitbritain.org/sites/default/files/vb-corporate/Documents-Library/documents/2015-1%20Level%20of%20interest%20among%20potential%20visitors%20to%20explore%20Britain%20in%20different%20ways.pdf (accessed August 1, 2020).

VisitBritain (undated) Gardens continue to attract the largest number of visitors. Available at: https://www.visitbritain.org/gardens-continue-attract-largest-numbers-visitors (accessed June 30, 2020).

VisitScotland (2017a) Latest Statistics. Available at: https://www.visitscotland.org/research-insights/about-our-industry/statistics (accessed June 6, 2020).

VisitScotland (2017b) Scotland Visitor Survey 2015 & 2016. Available at: https://www.visitscotland.org/binaries/content/assets/dot-org/pdf/research-papers/scotland-visitor-survey-2015-16-full.pdf (accessed July 22, 2020).

Vitruvian Planning (2018) Bown Crossing Branch Library Health Impact Assessment. A report to the City of Boise Idaho. Available at: https://www.boisepubliclibrary.org/media/9616/BownLibrary-Boise-HIA-DRAFT.pdf (accessed July 30, 2020).

Waimea Valley (Oahu, Hawaii). http://www.waimeavalleyhi.com/ (accessed August 10, 2020).

Wanhill, S. (1994) The measurement of tourist income multipliers. *Tourism Management* 15(4), 281–283.

Waterton, E. and Watson, S. (2014) *The Semiotics of Heritage Tourism*. Channel View Publications, Bristol, UK.

Wikipedia (n.d.) Happiness. Available at: http://en.wikipedia.org/wiki/Happiness (accessed May 26, 2014).

Williams, S.J., Jones, J.P.G., Gibbons, J.M. and Clubbe, C. (2015) Botanic gardens can positively influence visitors' environmental attitudes. *Biodiversity and Conservation* 24, 1609–1620. DOI: 10.1007/s10531-015-0879-7

Willis, C.K. (compiler) (2015) *Gardens for the Nation 1994–2014: Serving and Supporting South Africa's Social and Economic Development for 20 Years*. SANBI Biodiversity Series No. 27. South African National Biodiversity Institute, Pretoria, South Africa.

Willis, C. (2019) Draft National Botanical Garden Expansion Strategy 2019–2030. *Government Gazette* No. 42893, 82–113. Available at: https://cer.org.za/wp-content/uploads/2019/11/Draft-National-Botanical-Garden-Expansion-Strategy.pdf (accessed August 1, 2020).

Willis, C.K. and Mutshinyalo, T. (n.d.) Action Plan for the Kwelera National Botanical Garden. Unpublished report. South African National Biodiversity Institute, Pretoria, South Africa.

Wilson, E.O. (1984) *Biophilia*. Harvard University Press, Cambridge, Massachusetts.

Woods, C. (2018) *Gardenlust, A Botanical Tour of the World's Best New Gardens*. Timber Press, Portland, Oregon.

World Commission on Environment and Development (1987) *Our Common Future*. Oxford University Press, Oxford.

Wyse Jackson, P.S. and Sutherland, L.A. (2000) *International Agenda for Botanic Gardens in Conservation*. Botanic Gardens Conservation International, London.

Wysocki, C.J. and Preti, G. (2002) Human pheromones; oxymoron, marketing, maya or meaningful messages? *ChemoSense* 5, 1–11.

Wysocki, C.J and Preti, G. (2004) Facts, fallacies, fears and frustrations with human pheromones. *The Anatomical Record* 281A, 1201–1211.

Xiamen Botanical Garden (Xiamen, China). https://tools.bgci.org/garden.php?id=1036 (accessed August 10, 2020).

Xishuangbanna Tropical Botanical Garden, CAS (Menglun, China). https://tools.bgci.org/garden.php?id=338 (accessed August 10, 2020).

Zaccheus, M. (2015) Singapore Botanic Gardens Clinches Prestigious UNESCO World Heritage Site Status. *The Straits Times*, July 4. Available at: https://www.straitstimes.com/singapore/singapore-botanic-gardens-clinches-prestigious-unesco-world-heritage-site-status (accessed July 22, 2020).

Zamani-Farahani, H. and Fox, D.H. (2018) The contribution of rose and rosewater tourism and festival to the destination image. *Event Management* 22, 541–554.

Zar, H. (2016) Liberating Lyndhurst from the tyranny of the period of significance. *Forum Journal* 30(3), 27–36.

Index

Note: Page numbers in **bold** type refer to **figures**
Page numbers in *italic* type refer to *tables*
Page numbers followed by 'n' or 'nn' refer to notes